MILK AND HONEY

Inside Technology

Edited by Wiebe E. Bijker and Rebecca Slayton

A list of books in the series appears at the back of the book.

MILK AND HONEY

TECHNOLOGIES OF PLENTY IN THE MAKING OF A HOLY LAND

TAMAR NOVICK

The MIT Press
Cambridge, Massachusetts
London, England

The MIT Press would like to thank the anonymous peer reviewers who provided comments on drafts of this book. The generous work of academic experts is essential for establishing the authority and quality of our publications. We acknowledge with gratitude the contributions of these otherwise uncredited readers.

This book was set in ITC Stone Serif Std and ITC Stone Sans Std by New Best-set Typesetters Ltd. Printed and bound in the United States of America.

Library of Congress Cataloging-in-Publication Data

Names: Novick, Tamar, author.
Title: Milk and honey : technologies of plenty in the making of a holy land / Tamar Novick.
Description: Cambridge, Massachusetts : The MIT Press, 2023. | Series: Inside technology | Includes bibliographical references and index.
Identifiers: LCCN 2022033535 (print) | LCCN 2022033536 (ebook) | ISBN 9780262039079 (paperback) | ISBN 9780262374569 (epub) | ISBN 9780262374552 (pdf)
Subjects: LCSH: Agriculture—Palestine. | Agriculture—Israel. | Agricultural innovations—Palestine. | Agricultural innovations—Israel. | Agriculture—Religious aspects—Judaism. | Land settlement—Palestine. | Land settlement—Israel.
Classification: LCC S471.P3 N68 2023 (print) | LCC S471.P3 (ebook) | DDC 338.1095694—dc23/eng/20220822
LC record available at https://lccn.loc.gov/2022033535
LC ebook record available at https://lccn.loc.gov/2022033536

10 9 8 7 6 5 4 3 2 1

For my parents, Tova and Ilan

CONTENTS

ACKNOWLEDGMENTS

This is a book about plenty. Its conception and birth have depended on the support, kindness, enthusiasm, and friendship of plenty of people. I am indebted to them all and feel incredibly lucky that they agreed to join my journey.

This project began as a PhD dissertation that I developed as a graduate student at the Department of History and Sociology of Science at the University of Pennsylvania. I am grateful for the support of the Benjamin Franklin Fellowship and the School of Arts and Science's Dissertation Completion Fellowship. The department's community has been a source of knowledge, interest, and discoveries. Robert Aronowitz, my adviser and mentor, has been a fountain of support, encouragement, and care throughout the years. I am grateful not only for his kindness, patience, and reassurance but also for everything I learned from him about language, ideas, and humor, and reminding me about what matters in life. I thank my dissertation committee members, Heather Sharkey and John Tresch, for their support, availability, and thoughts. I extend my appreciation to Etienne Benson, Ruth Schwartz Cowan, and Rob Kohler for their generosity, detailed feedback, and faith in my work. I offer thanks to my teachers, David Barnes, Nathan Ensmenger, Steve Feierman, Riki Kuklick, Beth Linker, Julie Livingston, Projit Mukharji, Adriana Petryna, and Adelheid Voskuhl, and my fellow graduate students, Peter Collopy, Deanna Day, Erica Dwyer, Andrew Hogan, Elaine LaFay, Jessica Martucci, Samantha Muka, Jason Oaks (who taught me about caring for bees), Brittany Shields, and Kristoffer Whitney. I also thank Rachel Elder, Marissa Mika, and Joanna Radin for being good friends and readers.

X

The dissertation started becoming a book during the year I spent as a postdoctoral fellow at the Edmond J. Safra Center for Ethics at the Tel Aviv University, and I am grateful for its financial support. The center's fellows and research groups' participants challenged me in thinking outside my disciplinary comfort zone. I am particularly thankful to Irit Ballas, Hagai Boaz, Lin Chalozin-Dovrat, and Shai Lavi.

The Max Planck Institute for the History of Science in Berlin became my home, point of reference, and the place where I finished crafting the book. I am grateful for the institutional and financial support I received during my years at the institute, opportunities that it has offered, and wonderful people I was fortunate to meet. I thank Dagmar Schäfer for her encouragement and support. I extend my appreciation to the institute's library team for its endless help in making my research possible. Incredible people became part of my everyday life and thinking processes. Specifically, I thank my colleagues and friends Emily Brock, Esther Chen, Shih-Pei Chen, Alina-Sandra Cucu, Sebastian Felten, Wilko von Hardenberg, Stephanie Hood, Robert Kett, Lisa Onaga, Ohad Parnes, Martina Schlünder, Martina Siebert, and Mårten Söderblom Saarela. A special thank you to the participants in my working group, Out of Place, Out of Time, with whom I discussed animal reproduction, labor, and intimacy. I am grateful to Lucy Beech, Agata Kowalewska, Maria Pirogovskaya, Gabriel Rosenberg, Lukas Rieppel, Raphael Schwere, Shira Shmuely, Marianna Szczygielska, and Rebecca Woods. I also thank Sarah Blacker, Emily Brownell, BuYun Chen, and Noa Hegesh for their friendship and critical thinking.

Many friends and colleagues helped me think about this project and supplied invaluable feedback. I thank Alexander Shopov and Avner Wishnitzer for guiding me through the Ottoman Empire. I am grateful to Shira Wilkof for being an excellent and supportive reader and friend. Hila Vardi has been a wonderful listener and great reader. I also offer my appreciation to Gabi Admon-Rick, Samer Alatout, Gadi Algazi, Yoav Alon, Shukri Araf, Tal Arbel, On Barak, Soha Bayoumi, Erela Teharlev Ben-Shachar, Nimrod Ben-Zeev, Matan Boord, Irus Braverman, Başak Can, Nadav Davidovitch, Samuel Dolbee, Jessica Eary, Elik Elhanan, Tamar El-Or, Shai Enoch, Elia Etkin, Basma Fahoum, Talia Fried, Naama Gershy, Efrat Gilad, Snait Gissis, Motti Golani, Natalia Gotkowski, Rafi Grosglik, Yifat Gutman, Dotan Halevy, Ayala Hendin, Martin Hershenzon, Alma Igra, Yanay Israeli, Attara Ityel, the late Shaul Katz, Liat Kozma, Hagit Krik, Maya Levi, Kerry Levy,

Inna Leykin, Zachary Lockman, Itamar Mann, Jesse Olszynko-Gryn, Hagar Ophir, Dan Rabinowitz, Ahmed Ragab, Aviva Rot, Avraham Rot, Tiago Saraiva, David Schorr, Dmitri Shumsky, Dan Tsahor, Dora Vargha, Rotem Varon, Sigrid Vertommen, Simon Werrett, Rakefet Zalashik, Yahav Zohar, and all those I may have accidentally forgotten.

This book would also not have been possible without the help I received from various archives and research centers along with the generosity of my interviewees. I thank Arthur Kiron at the Herbert D. Katz Center for Judaic Studies at the University of Pennsylvania; the staff of the Historical Society of Philadelphia; the staff of the Widener Library at Harvard University; Sabine Dankwerts and Simon Renkert from the Domäne Dahlem Archive; the staff of the Beit-Rishonim Archive in Ness Ziona; Arieh Shadar at the archives of the Dairy Cattle Breeders Association in Kibbutz Yifat; Dalia Omer, Yosepha Pecher, and Yehudit Sedmy at the Kibbutz Mizra' Archives; the staff at the Kibbutz Merhavia Archives; Dudu Amitai, Yuval Dani'eli, and the rest of the staff at the Hashomer Hatzair Archives, Yad Ya'ari, Givat-Haviva; and the staff at the Kibbutz Yotveta Archives. I am grateful to Gady Alcalay, Hai Zumberg, and Ze'ev Zumberg for the tour at the cowshed at Kibbutz Yotvata. I also thank the staff at the Lavon Institute for the Labor Movement in Tel Aviv, Israel State Archives in Jerusalem, Central Zionist Archives in Jerusalem, American Jewish Joint Distribution Committee's Malben Archives in Jerusalem, and Hebrew University of Jerusalem Archives. And I am extremely grateful for what I learned from talking to Tzach Glasser of Ramat Hanadiv, Ze'ev Harari, Dorit Kababia and Yosef Karasso of the Sheep and Goats Section at the Ministry of Agriculture; Bruno Lunenfeld; Nir Mann; Avi Perevolotsky of the Volcani Center for Agricultural Research; Dudi and Ruti Silverman; Irit Sulman; and lastly, the late shepherd Lotek Etsion.

Conferences and workshops motivated me to test and develop my ideas. I am grateful for the conference audiences at meetings of the Society for the History of Technology, Society for Social Studies of Science, American Society for Environmental History, European Society for Environmental History, Israeli Society for the History and Philosophy of Science, and Israeli Anthropological Association. I am particularly thankful for the generous feedback I received during workshops that included On the Move: The Middle East and the "First Modern Globalization" at the Hebrew University of Jerusalem, Taking Animals Apart: Exploring Interspecies Enmeshment in a

Biotechnological Era at the Holtz Center for Science and Technology Studies at the University of Wisconsin, Science, Space, and the Environment at the Rachel Carson Center for Environment and Society, and Soil, Flesh and Flows: Environmental Temporalities and Expertise in the Middle East at the Center for Middle East Studies at Harvard University. I shared parts of this work with members of different research groups and seminars over the years, and benefited tremendously from the feedback, conversations, and suggestions. These groups and seminars include the Israeli Forum for the Study of the British Mandate, Science, Civic Legitimacy, and Local Knowledge research group at the Edmond J. Safra Center for Ethics, Laboratory for the Study of Environment and Society at Tel Aviv University, Generation to Reproduction seminar at the University of Cambridge, Colloquium for the History of Knowledge at Humboldt University, and Department III Colloquium Series at the Max Planck Institute for the History of Science. I thank everyone who asked questions and challenged my thinking.

Finally, words are insufficient to describe my gratitude to my family members for their support, generosity, and wisdom. I thank my three beloved sisters—Ruth, Yael, and Adi—and their families. I thank my mother and father, Tova and Ilan, for their endless encouragement and sparking my curiosity for the unknown. I thank Bashir for his heart of gold, and our son, Amir, who arrived in this world during the process of this book. They have taught me so much about love, life, and knowledge.

INTRODUCTION: A LAND FLOWING WITH MILK AND HONEY

There is an excited call from the girl at the machine. She turns on the tap and the first honey flows into the pail; deep, rich, amber fluid. Honey from the land which is flowing with milk and honey.

—Dorothy Kahn, "Flowing with Honey: How It's Done in a Jewish Settlement" (1938)

It was Zakia from Umm Djouni, a dwindling Arab village near the Sea of Galilee, who taught her how to make milk flow. Miriam Baratz, a young Jewish settler who arrived in Palestine four years prior to that transformative event in 1911, was determined to go against her parents' will in Russia and the demands of her fellow male settlers; she wanted to work the land. They used to call her "the wild goat" for this kind of stubbornness. The small group had recently settled in Umm Djouni, not long after Ottoman authorities allowed land purchases in Palestine, and as the village moved between Persian and Jewish hands.[1] They began forming Deganya, later known as the first communal agricultural settlement. Their single cow, which they named "First" (*Rishona*), was considered difficult, but Miriam was adamant that she could manage her despite having no relevant experience. She turned to her skilled neighbor who had "a good brain and a good heart."[2] The two women met secretly at night, and Zakia guided her through; success depended on a combination of bodily techniques and husbanding intimacy. Zakia let Miriam wear her blue gown, and they rehearsed Arabic songs considered necessary for the compliance of the animal. Milk finally flowed.[3] It was the beginning of a life in the cowshed for Miriam, who became a leading authority among the growing Jewish settler community. She nursed and raised her seven children in the shed while tending to the cows, and trained many generations of milk producers. A few years after that nocturnal female interspecies encounter, experts secured the bond between milk, governance, and settlement with publications such as *The Dairy Industry as the Basis for Colonisation in Palestine*, and Zakia disappeared from the historical record.[4]

With the stabilization of the British rule after World War I, milk became the pillar of settlement, its quantities a marker of success, and the bodies of cows sites of technological manipulation. Since the continuous production of milk depends on annual pregnancies, success was not limited to the crop of the body but also was tied to the fruit of the womb. Furthermore, this didn't only concern cows. As Miriam demonstrated with cows and through her own bodily labor, agricultural production and female reproduction were deeply entangled in the demographic transformation of Palestine and shaping changes in the land. The blending of progeny and flow of milk had a strong public face, as champion cows—producers of both milk and offspring—were lionized, and mothers of many children were prized; it reached new heights half a century later when the so-called Israeli cow was announced as a global milk-producing champion, and Israeli women and cows became global leaders in the consumption of artificial reproduction technologies.

The attention to milk production along with shifting demographic trends was not limited to the settler population but also nurtured by the shifting governing powers. The investment in the dairy industry was, in fact, part of broader Ottoman and British imperial efforts to enhance the implementation and use of agricultural technologies in a changing global economy.[5] It fit neatly with the British agenda of development and separatism.[6] During the interwar period, moreover, milk became an important material of concern for questions of demography, nutrition, and health far beyond the territories under British rule, and the investment in cattle breeding emerged as a central element across colonial regimes.[7] This made the cow a useful analogy for criticizing colonial governance and the systemic impoverishment of the people under its rule; as one Palestinian Arabic newspaper put it in 1930, "There is no milk left in this cow."[8] Indeed, such emphasis on milk production as the backbone of the settler colonial project was long-lasting and particular to Palestine; it was part of explicit efforts to literally create a land flowing with milk and honey.[9]

European settlers and the governing regime interpreted the land according to a widely accepted narrative of decline. The idea that a once-plentiful and green land became desolate was imperative to colonial rules across North Africa and the Middle East, as were the attempts to bring back plenitude through afforestation as well as the struggle against desertification and grazing goats.[10] Particular to governing Palestine was the belief that this past plenitude—of a land believed to be that of the Bible—was characterized by milk and honey. In 1942, Gilbert Noel Sale, who became the conservator of forests in British Palestine after holding posts in colonial Cyprus and Mauritania, detailed "the history of erosion in Palestine" to an audience of experts:

> The land was originally covered with a forest which varied in height and composition. . . . Multitudes of flowers were visited by the bees which provided wild honey, one of the foods of early man. In due course, as we know from ancient literature, man evolved from the stages of hunting and honey collecting, and began to keep domestic animals. The changes in the vegetation and in the condition of the land dated from the time when the country was flowing with milk and well as honey. At first, no doubt, little damage was caused by small flocks of goats and sheep which wandered in the great forests . . . and it was not until man became more completely master of his environment that he enlarged his flocks to dangerous proportions. . . . [S]ubsequently, the invasions of less civilized races, unversed in the agricultural arts, led to the neglect of the terraces, which rapidly decayed.[11]

Milk and honey were Palestine's natural condition, according to this common view, and the behavior of latecomer people and animals—interpreted as unnatural and harmful to the land—disrupted its balance and were its ultimate source of decay. "Palestine is a natural garden," summarized Sale, who was also poet, "and must be restored to its original condition."[12]

This book hones in on such plans for restoration, and the technological means state experts and settlers employed to materialize a religious idea of the land, even when those made little economic or environmental sense.[13] From the turn of the twentieth century, Christian and Jewish settlers as well as the changing governing powers utilized a variety of tools and techniques in order to demonstrate that Palestine could literally flow with milk and honey, and eliminate what they understood to be the causes of damage to the land. Put differently, these were both technopolitical and envirotechnical plans.[14] Such plans to realize a biblical metaphor, I argue further, were carried out on a pragmatic level through the bodies of animals and people. In 1921, for example, the US consul in Jerusalem reported home about the state of beekeeping, noting that the question of "whether or not Palestine may literally become a land flowing with milk and honey is now being tested in a practical and a commercial manner."[15] Similarly, journalist and settler Dorothy Kahn summarized her experiences in beekeeping in 1938 with the practical title of "Flowing with Honey: How It's Done in a Jewish Settlement."[16] Indeed, as we will see, different kinds of political, social, and religious powers came together in using technological means for re-creating the Holy Land in modern Palestine.

Demonstrating plenty depended on human and animal, often female, bodily labor. As reflected in the logic of Sale's historical narrative, this process was also frequently entangled with disregard for, and the delegitimization and criminalization of, local Palestinian forms of life and knowledge. But as the story of Miriam and Zakia illustrates, this same process deeply depended on local expertise. These types of knowledge, such as the manner by which people tended their animals and made them produce, were frequently unrecognized, repeatedly erased from the historical record, and ultimately forgotten. This book engages with a specific kind of such "disabled histories," and as a result, particular kinds of appropriated knowledge—intimate knowledge—of living bodies along with their production and reproduction.[17]

Furthermore, an ongoing tension existed between intimate forms of knowledge that grounded the settler colonial plans and formal means of governing the land. Numbers were such a formal means, albeit certainly not new to the region. Some of the cardinal techniques of Ottoman governance included enumeration and registration for the purpose of taxation. The accumulation of funds was, in turn, key for sustaining the empire. Beginning in the mid-nineteenth century, with growing European involvement in the region, surveying and mapping joined the intensifying efforts to quantify Palestine. Numerical data, which unlike intimate knowledge, allows for growing levels of abstraction, standardization, and universalization—indeed, the foundations of modern science—was essential for knowing and controlling the land, and ultimately, regulating everyday life under British and Israeli rules.[18]

Numbers had another purpose: they allowed settlers and state experts to prove that the land was becoming, again, a land of plenty. In this sense, counting and measuring were pivotal "technologies of plenty." These along with other tools and methods of demonstrating plenty form the focus of *Milk and Honey*. Among them are beehives, lists of livestock, breeding practices, pregnancy tests, and fertility treatments, but also bodily and sensorial techniques. Although settlers and state experts such as Miriam and Sale make frequent appearances in the book, its main subject is other creatures—ones that produce milk, honey, and offspring, and embodied or interfered with the plans to restore the land—water buffalo, bees, goats, sheep, cows, and finally, women.

A DESIRE FOR A HOLY LAND

Travelers to Palestine have long been occupied with describing the land and its religious traction. Ottoman officials, such as Āşik Mehmed, Kātib Çelebi, and Evliyā Çelebi, who traveled the land during the sixteenth and seventeenth centuries, supplied their impression of the landscape, people, and animals. While earlier travelers, following medieval cosmographic works, focused on comparing classic Arabic texts with their own observations, later depictions relied on a combination of personal impressions and the collection of local testimonies.[19] The Napoleonic and Crimean Wars that were a watershed in global politics and the economy ultimately enabled European powers access to Palestine along with gradual control over many of its territories. The professionalization and growing success of biblical studies among midcentury European intellectuals, in conjunction with archaeological findings from the ancient civilizations of the Near East, contributed to changing perceptions of religious texts.[20] These combined political, economic, and intellectual processes resulted in increasing numbers of Europeans and Americans in Palestine. The scope of traveler accounts grew dramatically in the second half of the nineteenth century, along with lengthy descriptions and analyses of the land by researchers, imperial agents, and settlers. Their view of the Bible as a record of historical events shaped their experiences; travelers and settlers were carrying the Bible in their hands, seeking to find the Holy Land there.[21]

What Europeans found in Palestine, however, was dramatically different from what they expected. Both Christians and Jews were overwhelmed with the desert that stood between what they hoped to find and the unfamiliar land. A British traveler to Palestine wrote in 1882, "My first strong impression, and, I may say, my last, on beholding Palestine was one of astonishment. Can this be that glory of all lands—that Promised Land— the land flowing with milk and honey? No! Surely not. . . . I had pictured fertile plains and dewy meads . . . cultivated lands bringing forth luxuriant crops almost spontaneously. . . . Palestine, of all countries, is now desolate, barren, and accursed."[22] European comers to Palestine wanted to find a Holy Land, but what they encountered was a place that seemed worse than profane.

The biblical phrase "a land flowing with milk and honey" appears many times in the scriptures, and has usually been interpreted as a metaphor for abundance.[23] Many generations of Christians and Jews used it as a way to imagine the Holy Land. For centuries, for example, Christian Europeans hymned—in both Latin and English—"Jerusalem the Golden with Milk and Honey Blest. . . . I know not, O I know not, what joys await us there, what radiance of glory, what bliss beyond compare."[24] Yet with the growth of European presence in Palestine, this metaphor became a powerful tool for demonstrating the gap between the imagined and the real, and exposed ignorance about local forms of life, production, and expertise. Newcomers to Palestine commonly used the phrase "a land flowing with milk and honey" to dramatize their sense of disappointment: "Is this the land of my fathers? The land that is said to flow of milk and honey?" asked a discontented Silesian Jewish traveler to Palestine in 1838.[25] "Jerusalem the Golden with Milk and Honey Blest, Where is that Milk and Honey? It seemed to have 'gone West'" sang British troops as they took control over Palestine in 1917.[26] But the use of the phrase did not end with that bitter disappointment. The image of a plentiful land, a land flowing with milk and honey, became idealized, emerging as an organizing principle for the changing political regimes and growing European settlement in Palestine.

This phenomenon of understanding and treating a land according to previous expectations was not unique to Europeans in Palestine but instead was prevalent across colonial contexts. In *Changes in the Land*, for instance, historian William Cronon argues that the descriptions of the first European settlers in New England reflect both contemporary environments and their own ideological biases. The way settlers in the Americas viewed the land was heavily influenced by the potential profit to be made by circulating its resources in European markets.[27] The expectations of Europeans in Palestine were manifested through a consistent search for plentiful land, a Holy Land.[28] The use of this specific metaphor—"flowing with milk and honey"—to portray a land was not unique to Palestine/Israel either.[29] Indeed, this biblical phrase and other expressions of fecundity have frequently been used to describe fertile and plentiful environments of other "newfound lands," such as the Americas and Oceania.[30] Adopting the Bible as a historical document, however, Europeans in Palestine believed that the land of Palestine was *the* land depicted in the Bible, and as such, it should literally be full of milk and honey.

The theme of decline—the idea that the land used to prosper in biblical times, but had since decayed—was widely discussed and debated in European interpretations of the land.[31] Numerous state experts and settlers occupied themselves with analyzing ancient prosperity and the process of impoverishment. Furthermore, this debate regarding the ancient past encapsulated grander tensions of European colonialism in the Middle East. In their perception of the Middle East as the "cradle of civilization," Europeans were torn between their belief in a glorious past that they considered their own and their faith in the tools of modernity. Some suggested that the land of the Bible was as bountiful as a land could be; others asserted that while the process of environmental degradation was evident, the land of the Bible was naturally meager. For both approaches, the main motivation for controlling the land of Palestine was the belief that unusual, unique things had happened there in the past. Yet in order to justify seizing control and transforming Palestine, the future had to look brighter than this past. To the governing powers and for the settlers, the land, as part of this paradigm, should become extraordinarily plentiful and its creatures extremely productive.

TECHNOLOGIES OF PLENTY

Three main threads interlace throughout this book: first, the ways agricultural technologies were utilized to materialize a particular, religious notion of the past; second, intimate forms of knowledge and bodily labor—production and reproduction—as the main spheres in which this process took place; and finally, how the bodily, political, and environmental realms intertwined in the transformation of Palestine.

Technology, religious fervor, body labor, and political ecology fuse in this story. A wooden movable beehive became a tool for demonstrating that land was changeable; early Christian settlers considered the flow of honey that this beehive allowed as a literal expression of the land's revival. The growth of the dairy cow population and successful management of its fertility were similarly seen as verification of the land's exceptionality; the crosses of European and Middle Eastern cattle breeds, which broke milk production records both locally and globally, were viewed as the ultimate proof of the power of technology in erecting plenty, and considered a justification for the growing European intervention in the region. Problems with the production of plenty were managed with scientific means: when cows, sheep, and women failed to reproduce, settler farmers, veterinarians, gynecologists, and endocrinologists came together to find solutions to infertility across species divides. Other challenges to the attempts to "make the desert bloom" necessitated the rhetoric and tools of ecology and demography. When the behavior of water buffalo came to interfere with intensive agriculture, their long-standing valuation dissipated, and movement restrictions resulted in their eventual disappearance. When goats defied newly established borders between private and state land, they came to be seen as enemies of nature and the state. Large-scale British and Zionist tree-planting efforts—understood to be a project of reforestation—were threatened by the appetite of black, herding goats. Their feeding needs won them, along with their Palestinian Arab owners, the title of "the creators of deserts."

Such examples challenge our assumptions about the relation between religion and modern science and technology, revealing their nonbinary nature. Much work is dedicated to showing how science emerged from religious ideas and the manner in which scientific thinking grew from religious institutions, ranging from ancient times and until the early modern period.[32] But when it comes to late modern science, the majority of the scholarship describes a world where science and religion clash.[33] It is also a world in which scientific and technological superiority replaced religion as a tool of governance.[34] Some recent work has contested this notion of discrepancy, demonstrating how late modern scientific thought and technological practice have helped appease the tensions inherent in nationalism—tensions between narratives of mystical pasts and utopian, rational futures.[35] Yet the tendency to place religion and modernity in opposition along with the belief in the secularizing power of science remain dominant in studies of Palestine/Israel; Zionism is repeatedly understood to be a national and

FIGURE I.1

"Modern science is taught in the home of ancient religions," an image from a lengthy article published in the *National Geographic Magazine* depicting the transformation of life, practice, and landscapes in Palestine under British rule. *Source:* Edward Keith-Roach, "Changing Palestine," *National Geographic Magazine*, April 1934, 510; photo by S. Kaplansky.

settlement movement that raised the banner of modernity, secularism, and technocracy.[36] The case studies that follow reveal the ways in which technology did not replace religion as a colonial device but instead was blended with aspirations to salvage the land, and how this blending became crucial for seizing control over lands and people.[37]

These efforts to utilize technology for re-creating plenty were exercised primarily through bodywork and the formation of life, both of humans and other animals.[38] Bodies, in other words, were the main sites for exercising this change. One was the crossbred body that Jewish settlers called the "Hebrew cow," and that labored to produce a flow of measurable milk. Another was the body of the male Jewish settler, who was trained by Bedouin shepherds to move and use senses and voice in particular ways in order to communicate with as well as manage sheep. These bodily movements of shepherds and sheep across the Palestinian landscape allowed settlers to become Hebrews and hence connect to the practices of biblical leaders. By

paying attention to the body in relation to the environment, the book follows other histories of settlement and colonialism.[39] It describes the ways in
which bodies were used and manipulated in order to impregnate the land,
ultimately recounting a tale of producing new modes of becoming native.

An extensive body of work in the fields of colonial studies, feminist
theory, and anthropology attends to intimate relations in order to portray,
respectively, the colonial situation and the essence of power, the global,
and the universal.[40] *Milk and Honey* similarly pays attention to intimacy,
extends its application to the nonhuman, and demonstrates how managing the functionality, labor, and fertility of (primarily female) bodies relied
on knowledge in proximity, and a mitigation between individual bodies
and generalizable data. Through this focus on intimate knowledge, the
book engages with the sensibility to particularities and the local in recent
science and technology studies scholarship, and goes further to align them
with large-scale historical processes such as settlement and colonialism. It
shows how managing the bodies of both animals and humans was foundational to such processes. Following historian Tiago Saraiva's exploration of
how organisms embodied fascist regimes, which he defines as "a study in
fascist ontology," this book supplies a study in settler colonial ontology.[41]
Without paying attention to knowledge of the intimate, I contend, we cannot understand technopolitics; the most profound forms of technopolitics
are biopolitical.[42]

Creating a land of plenty in Palestine was a more-than-human process
at its core.[43] Animals have long been central to colonial and environmental
histories, and gradually, also key protagonists in histories of science and
knowledge of nature.[44] This book builds on these perspectives as well as
recent sociological and anthropological attempts to consider animals as and
in relation to laborers.[45] It then bridges such approaches from animal studies with the history of the body to show that labor, production, and reproduction of both humans and animals were considered as well as managed
in conjunction with one another within this particular political context.
By tracing the joint understanding and management of the reproduction
of animals and humans, this book also offers a prehistory to the so-called
fertility revolution in Israel, while undermining its strictly human, cultural,
and socioeconomic scholarly reasoning.[46]

Cutting across three political regimes—the late Ottoman rule (~1880–
1917), British rule (1917–1948), and the early Israeli state (from 1948

on)—*Milk and Honey* explores the various ways that state experts and set-tlers understood the land, and highlights how configuring bodies and the environment intertwined with governance and the construction of settler society.[47] As an environmental history, this book analyzes European efforts at political domination as well as ecological control and manipulation as deeply interrelated. It pays particular attention to the ways in which con-quest, governance, and settlement were entangled with dramatic biotic transformation, and to ideas about the environment, or "environmental imaginaries," that were used to justify colonial expansion.[48] Unlike studies of agricultural theory and institutions, which have dominated the schol-arship on Palestine/Israel and Zionism, this work focuses on skill, prac-tice, and embodiment.[49] *Milk and Honey* deals with the labor and value of water buffalo, goats, and bees. It tells the stories of the holy bee queen that crossed the ocean, Shatra the rebellious sheep, Stavit the prolific and celeb-rity cow that outlived her productive life, and even Rutie the horse, whose bodily fluids were used to appease anxieties about infertility among other female, human and nonhuman, creatures. By so doing, this work exposes how large-scale environmental and political changes were defined through the labor, physicality, and limitations of individual bodies, and illustrates how the presence, knowledge, and practices of Palestinian people shaped European governance and settlement in Palestine/Israel.

CHRONOLOGY AND STRUCTURE

The global political, economic, and demographic transformations of the nineteenth century left their mark on late Ottoman Palestine. Midcentury, ongoing changes to landownership and use received their formal legal stamp as part of the Ottoman program of reforms known as the *Tanzimat*, and the possibility of private property gradually gained prominence at the expense of the communal land tenure system (*mushā*).[50] As the case of the Umm Djouni village demonstrated, foreign money became an important player in Palestine. By the 1880s, not only European and US travelers and researchers but also settlers became common to the land. By studying, writ-ing about, investing, and settling in Palestine, these various groups contrib-uted to making Palestine legible to the West.[51]

World War I brought an end to the Ottoman Empire; in 1917, Brit-ain seized military control over Palestine, and starting in 1922, governed according to the contemporary mandate system. This period transformed the settlement patterns, economic structures, and environmental policies of the area. The dramatic growth of European (predominantly Zionist) settlements in Palestine, grossly encouraged by the British rule, sparked great tensions between the Jewish and Arab populations, and between those and the governing rule, on methods of the ownership and use of the land. Escalating tensions turned violent in the 1920s, and particularly intensified from 1936 until 1939, in a series of events remembered as the Arab Revolt.[52] The 1948 war shifted power structures and population composition in Palestine, of which major parts became the State of Israel. During the war, many hundreds of thousands of Palestinian Arabs were forced into exile (740,000 people according to recent estimates), and only a small number were able to stay and live under the new military rule (for Palestinians only), which lasted until 1966.[53] In those years, more than a million Jews immigrated to Israel from Europe and the Arab world, settling in Palestinian houses or other dwellings in cities, and joining existing or new agricultural settlements and peripheral towns.

Milk and Honey is organized around the production of milk, honey, and offspring, gradually moving from the late Ottoman period, through the British mandate years, to the early Israeli state, with a focus on the period between 1880 and 1960. The book is primarily based on what can been characterized as the "colonial archive" and "settler archive," and fuses a combination of sources: state-produced documents such as legislation, taxation, and administration records, correspondence between state experts and bureaucrats, petitions sent to state officials, scientific publications, newspaper items, personal records, poems, fables, memoirs, and oral histories. The great majority of these sources were written, published, or recorded in Hebrew and English, others were in Arabic, and a small portion were in German or French. Following anthropologist and theorist Ann Laura Stoler, I analyze these diverse and complementary types of sources preserved at the colonial and settler archives by reading along the archival grain.[54] The curatorship and presentation of this diverse collection is key for proving that the efforts to demonstrate plenty were widely practiced, and brought together a variegated group of participants—different British and Israel state experts and stakeholders as well as different Christian and Jewish settlers. By doing so, the book ultimately attempts to describe the settler colonial condition.

The book begins by tracing water buffalo throughout Ottoman Palestine by way of supplying a prehistory of plenty. I use Ottoman taxation records as well as ethnographic evidence and oral histories in order to estimate changes in the number, use, and importance of milk-producing buffalo in Palestine, and recount the process by which this once-prevalent and valuable animal was eventually removed from the landscape by the mid-twentieth century.

Chapter 1 then analyzes the emergence of "modern beekeeping" in Palestine at the turn of the twentieth century and centers on the use of one technology: the movable frame beehive. It looks specifically at the story of an Alsatian Christian missionary family that utilized this wooden hive to transport numerous honeybees across country, and is based on family records and publications. European settlers used this technology and the movements it allowed in an effort to demonstrate not only the land's sanctity—in showing that it was literally "flowing with honey"—but also the power of Western interventions in transforming the "immovable East." With attention to changes in the lives of bees and production of honey, I examine how early European settlers and the shifting regimes interpreted the role and limitations of honey-making machines.

Chapter 2 focuses on the denunciation of the leading milk producer of the area: the black herding goat. It examines the process by which British and Israeli experts and state officials came to see the hungry goat as a threat to the revival of the land. As the records of the British Forestry Department and Israeli Agriculture Ministry at the Israel State Archives reveal, the tools of denunciation included counting, recording, measuring, and classifying, ultimately culminating in a plan to terminate these destructive creatures and replace them with prolific others. By analyzing a series of petitions sent by Palestinian goat owners to British and Israeli state officials, the chapter considers the professional and lay debates and actions that emerged as a result of these convictions along with new ways of organizing the land.

Knowledge of sheep herding is at the heart of chapter 3. It delves into Jewish settlers' attempts to revive a biblical practice and hence the land in the first half of the twentieth century. I examine two organizations of Jewish settlers with opposing ideas about practicing shepherding, and compare the types of knowledge and practices they considered valuable: either relying on the senses or employing numbers. As shepherds themselves became obsolete, however, the sheep milk economy was in the end unsuccessful.

Based primarily on the records of these organizations and their members, the chapter explores this once once-widespread agricultural and cultural phenomenon of shepherding as a way of considering the meanings of failure.

Chapter 4 explores dairy farming among Christian and Jewish settlers during the British mandate and early State of Israel. It revolves around an analysis of the records of the Dairy Cattle Breeders Association, and the invention and costly success of the Hebrew cow. I ask why the foreign bovine dairy industry was positioned at the center of the growing settler's agricultural economy and track the process of creating a high milk-yielding species, which was the result of multiple attempts at breeding local and European cows. As milk yield became the new way of measuring success, breeding practices interacted with and changed in response to environmental challenges and political upheavals. In dialogue with the "New Jew," the term utilized to explain the ideal Jewish male settler body, the chapter also introduces the term "New Jewess" to contend that producing a plentiful land depended on fertile female bodies.

Chapter 5 scrutinizes the problem of infertility, which threatened the existence of the entire settlement project. It focuses on a group of settler gynecologists and veterinarians, and their correspondence and collaboration with farmers in an attempt to deal with the reproductive limitations of the human and animal body. Through this work, the urine of mares and women—an abundant source of sex hormones—emerged as a savior substance and connected the farm to the clinic. This chapter utilizes laboratory and farm records as well as scientific publications addressed to the global scientific community in order to show the extent of efforts put into realizing plenty.

Taken together, the chapters offer a way of looking at the connections between the social order and transforming environmental and biological ones. Following the paths of production and dysfunction, the book ends with a discussion of the meanings and consequences of such images of entanglements. It proposes and supplies evidence—such as the stories of Methuselah, the lonely male palm tree that could not reproduce, and superior female bovine embryos—to show that many technologies of plenty are alive and well.

INTERLUDE: BYGONE BUFFALO AND LINGERING VALUE— A PREHISTORY OF PLENTY

The sun has risen, oh so very beautiful.
Come on. Let's go milk the water buffalo.
—A popular Arab song

I am a rare cock, I am a clever cock!
. . . for the water buffalo I got a bride.
—A Palestinian fable

Palestine was certainly flowing with milk and honey—at least as far as Ottoman records are concerned. From its inception, the people of Palestine experienced the Ottoman rule (1516–1917) through changing urban architecture, the construction of large-scale water and road infrastructures, and most notably, annual taxation. Taxation, which enabled the growth and maintenance of the empire, was initially based on a series of surveys that the government used in order to estimate and negotiate the structures of production and profit. This system, imposed on the predominantly rural population of Palestine, was dedicated to agriculture and the trade of its products. It consisted of few taxable categories, such crops, animals, and market activity.[1] While many animals were involved in and central to daily life and practice in Palestine, including donkeys, horses, camels, sheep, cows, and chickens, only three were taxed throughout nearly four hundred years of Ottoman rule: bees, goats, and water buffalo—the principle producers of milk and honey in the area.[2]

The last surviving early survey in the Arab provinces was taken in 1596–
1597, a time in which the Ottoman Empire was arguably prospering in
terms of population, production, and funds.[3] Hence its registers, which
demonstrate a knowledge of the land and its predicted value, supply us
with an image of plenty. They also mark a period of transformation: an
end to the era of the direct taxation system (*tapu resmi*) and a shift to the
tax farming system, in which local individuals were granted temporary (*ilt-
izam*) and later lifelong (*malikâne*) authority to collect tax. The new system
was based on an assessment of monetary value rather than on knowledge
by kind; the state no longer knew how many animals lived and produced
within its territories, but calculated their production's worth instead.[4] The
change, which resulted in a gradual decentralization of the power of the
state and the strengthening of local elites, aligned with larger political,
economic, and environmental challenges that unfolded during the seventeenth and eighteenth centuries.[5]

In Palestine, the southern part of the Syrian provinces that was often
considered marginal to imperial interests, local elites gradually shaped the
agricultural economy in relation to the land's diverse ecology and changing
tenure. A decades-long investment in growing cotton and olive trees for the
soap industry in the hinterlands as well as wheat and barley in the central
and southern regions was expanding in the seventeenth century in order
to satisfy regional demands.[6] The strengthening of the coastal plain and
its ports in the eighteenth and nineteenth centuries went hand in hand
with the expansion and sheer growth of Palestinian production as well as
Palestine's deeper integration into the global economy, as the cases of olive
oil, sesame seeds, and citrus fruits exemplify.[7] In addition to an Egyptian
occupation that was short-lived (1832–1840) but had a decisive impact,
the Ottoman reformative period of the nineteenth century, known as the
Tanzimat, saw dramatic changes to managing lands and people under its
rule. It also brought the survey system back to life, reviving the old empire-
wide means of knowing the land. Following regional military conquests,
European involvement in Palestine became apparent first through surveys,
mapping, and flora and fauna collection projects, then land purchasing
and private concessions over resources and parts of the country, and finally
settlement. An examination of the 1596–1597 survey therefore allows for a
look into the ways in which the Ottoman rule understood life in Palestine
on the brink of wide-scale changes.

Bees and goats were central to the livelihood of the great majority of villagers in Palestine during the Ottoman period.[8] Almost every village among the many hundreds counted in the late sixteenth century paid annual fees for beehives and goats, presumably often including sheep.[9] As we will see in the following chapters, milk and honey were produced seasonally, according to regional vegetation and reproduction patterns. Bees were raised in stable clay hives, and goats and sheep were herded across the Palestinian landscape. Goat milk and honey were brought to urban markets during production seasons, and unused milk was processed into dried cheese and butter. Beyond milk and honey, the people of Palestine utilized other bodily substances of animals: they ate the meat of sheep and goats on special occasions, used goatskins to carry water or churn butter, used goat hair for tents and manure for hut construction, and applied bee stings to wounds and used honey as medicine.[10] Goat and sheep manure were also central to maintaining the fertility of soil, connecting the presence and movements of these animals to the agricultural economy as a whole, and prominent crops such as wheat, barley, olives, cotton, vines, beans, vegetables, and watermelons.

The third animal recognized as important to Palestinian life and the economy through taxation records is the water buffalo, locally known as *Jamus/a*. According to the records of the 1596–1597 survey, the people of Palestine paid approximately 65,000 Ottoman aspers (*akçe*) for the buffalo.[11] Considering the known rate of 4 to 6 akçe for every water buffalo being milked, it seems that about 13,600 female water buffalo lived and produced in early Ottoman Palestine, along with an estimated population of about 200,000 people.[12] It is a striking number, which stands in stark contrast to the underdocumented and nearly forgotten history of this creature in Palestine.

It is assumed that water buffalo arrived in Palestine from South Asia with the Muslim conquest at the turn of the eighth century, but some archaeological evidence undermines this claim.[13] Some accounts hint at the increasing importance of the animal to the region; writing in the early fourteenth century, Mamluk bureaucrat Shihāb al-Dīn al-Nuwayrī, for example, described the use of water buffalo as in protecting against lions, tilling the soil, carrying loads, and pulling wagons in the lowlands and coastal areas of Syria. He noted that "the milk of the water buffalo is among the most delicious and richest of all types of milk," and spoke of the intimate relation established between the animals and their herders in Syria and Egypt, arguing that "the herders call each animal by its own name, which it recognizes when called to be milked."[14] Water buffalo became key to the regional economy during the Ottoman period. As historian Alan Mikhail demonstrates, female water buffalo were the single most valuable animal in Ottoman Egypt.[15] Indeed, according to late nineteenth- and twentieth-century writers, the main tenders and breeders of water buffalo had most likely arrived in Palestine from the surroundings of modern-day Sudan with the Egyptian conquest in the 1830s, possibly as slaves.[16] They settled in mud huts near water sources such as the Auja River and Kabara wetlands along the western coast, or Lake Huleh and the Jordan Valley in the northeast. In addition to keeping buffalo for milk and to plow, they made their living from papyrus mat making.[17] They were Bedouins, commonly referred to as the Ghawarneh tribe, of the lowest social and economic strata, and had experienced discrimination for their origins, living arrangements, and color.[18]

Based on the extent of the research, and considering that their tenders left little record, it is difficult to tell how water buffalo lived and have been managed in Palestine, and how that changed over time. But some folkloric evidence helps give a sense of their value. One example is a Palestinian children's fable, which was documented by two different ethnographers in the early twentieth century and features a trickster cock. At the beginning of the story, the cock finds a grain of corn. He brings it to a woman at a mill and asks her to grind it. After grinding, he asks for his grain back, thus forcing the woman to compensate him, which she does by giving him some flour to bake a loaf of bread. The cock then continues to manipulate people he meets by offering them the food he holds and then asking for it back after it has already been eaten. By this manner, the cock manages to replace the loaf with a bunch of green onions, the onions with a young sheep, the sheep with a camel, and the camel with a water buffalo. Finally, he tricks a big poor family into eating the buffalo. Unable to give the animal back on the cock's request, the head of the family offers one of his seven daughters instead. The cock then crows, "I am a rare cock, I am a clever cock! For the grain I got a loaf, and for the loaf I got a bunch of onions, and for the onions I got a kid, and for the kid I got a camel, and for the camel I got a water buffalo, and for the water buffalo I got a bride."[19] In this hierarchy of value, notwithstanding the outsmarting bird, water buffalo are positioned at the top of the plant and animal kingdom. The story illustrates not only that water buffalo were important, though. The value of water buffalo as depicted in the fable is, in fact, also comparable to that of human beings.

In addition to such accounts, oral histories confirm the significance of this animal to Palestinian Arabs in the early twentieth century. Warda al-Abdallah (born 1922), a Palestinian 1948 war refugee from the village of Mallaha near Lake Huleh, for instance, recounted the different practices revolving around water buffalo, noting their outstanding value and the superior quality of their milk.[20] While milk was the main crop drawn from the body of buffalo, hence determining the manner by which these bovines were raised and bred, buffalo were used in the fields too, and their meat became food in special circumstances, such as on the death of a close family member. Like in the case of goats, the skin of buffalo was made into water or butter containers, and their hair was used for mattress making. Muhammad Qasim Muhammad (born 1926), a herder, mat maker, and later refugee from Jahula, detailed the daily rhythm of moving the animals to the woods and water, and then back to the village. Water buffalo were easy to work with (in comparison to cows), he observed, and tended to wait to be picked up from the village's different owners in the morning.[21] Ahmad Ismail Dakhloul (born 1918), a refugee from the nearby village of al-Salihiyya, explained how water buffalo labored to plow and detailed the various kinds of dairy products made from their distinct, rich milk.[22]

By the late nineteenth century, water buffalo disappeared from Ottoman enumeration records and were marginal to the censuses of the British rule that followed (1917–1948).[23] According to British estimations, their numbers decreased substantially during the first half of the twentieth century, with 4,480 buffalo counted in 1942—about a third of the early Ottoman estimations.[24] The above depictions of value and use therefore describe a reality in which the number of water buffalo as well as their significance to the governing power were already substantially diminishing. It was value recognized in the midst of a process of obliteration.[25]

European and US travelers to Palestine, a vocal element in the growing multifaceted intervention in the land from the mid-nineteenth century on, were mesmerized by the appearance and size of water buffalo, and considered their presence a proof that Palestine was the abundant land of the Bible, a "land flowing with milk and honey." Writing about his visit to the Huleh plain, US missionary William McClure Thomson remarked in 1859 that its soil was "extremely fertile." Describing the life of the Ghawarneh people, he explained that they "make large quantities of butter from their herds of buffalo, and gather honey in abundance from their bees. The Huleh is, in fact, a perpetual pasture-field for cattle, and flowery paradise for bees. . . . Thus this plain still flows with milk and honey." Along with this tendency to equate Palestine with the scriptures, the figure of the buffalo carried different meanings. In his book, Thomson discussed at length the bodily features and habits of the buffalo, highlighting their dark color, monstrosity, and tendency to wallow in the water:

> Large herds of buffaloes lie under the covert of the reeds and willows of the many brooks which creep through this vast marsh, and we shall see them all day, as we ride round it, wallowing in the mire like gigantic swine. There are larger than other cattle in this region. Some of the bulls are indeed rough and monstrous fellows, with bones black, and hard "like bars of iron." With the aid of little Oriental hyperbole I can work up these buffaloes into very tolerable behemoth.[26]

A similar unfavorable understanding of the animal is reflected in the development plans and writings of British and Zionist experts in the early twentieth century. They, too, thought that the soil of the Huleh was fertile, but considered its body of water a swamp, a source of disease and stagnation that necessitated large-scale drainage.[27] Such efforts to reveal the Huleh's agricultural potential entailed dramatic landscape manipulation, which the behavior of water buffalo interrupted: "There is nothing more destructive to artificial channels than uncontrolled cattle," observed a group of British consulting engineers to a Zionist funding body in a report on the reclamation plan in 1936. "They trample down the banks to get to drinking water, [and] the buffalo likes to wallow and creates nasty muddy pools."[28] This perceived disturbance was translated into official restrictions on their grazing and use of water sources, which challenged formerly established ways of living with buffalo.[29] Water buffalo, wrote colonial veterinary officer I. Gillespie in 1943, "are confined to the Huleh area in the Galilee District where the presence of marsh land provides a suitable habitat. As reclamation of the region proceeds, the buffalo will become less important. For the time being they provide milk and traction for their Arab owners but if the area is drained it will become more economical to substitute them."[30] Lake Huleh's drainage was completed in 1958, under the State of Israel, and soon developed into an ecological crisis.[31] Water buffalo finally disappeared from their last habitat, official records, and public memory, along with their tenders, knowledge, and value.[32]

By the mid-twentieth century, the land and its creatures were radically transformed. The changing governing regimes and growing settler population, which adopted the notion of a land flowing with milk and honey as a plan for managing Palestine, chose other animals as the basis for its implementation. It is this long history of animal enumeration, use, and value as well as the relinquishment of water buffalo and their tenders that we should consider before delving into the story of *Milk and Honey*.

1 BIBLE, BEES, AND BOXES: TECHNOLOGIES OF MOVEMENT AND OBSTRUCTION

Like many European and North-American people, honeybees began to travel throughout Palestine in the last decades of the nineteenth century. While nothing is surprising about bee flight, the circumstances of the nineteenth century created a new type of bee movement. Beyond the local "Holy Land" tourism frenzy, bee travels were part of extensive exploration projects in all European colonies during this time.[1] One of those projects focused on finding "the ultimate race of bees"; bee species "supremacy" depended on the quality and amount of honey produced, their temper, and resistance to disease.[2] By the first decades of the twentieth century, the "Italian bee" had prominently established itself as the preferred breed in most parts of the globe, yet several other types of bees were thought to have powerful qualities, particularly the bees of the East, *apis dorsata* (see figure 1.1).

Until the late nineteenth century, bees in Palestine encountered a limited variety of flora—simply that which existed within a radius of about two miles of flight. Bees created honey during a short period of the year, and some of them lived in human-made hives, which were usually long, cylindrical, and made of clay. The bees built the honeycomb anew after every swarming season, since human hands had to break the hives in order to extract the honey. Honey was valued in Palestine as an article of food and was often thought to hold medicinal qualities; it was consumed locally and at times sold in nearby markets.

By the early twentieth century, the life of bees in Palestine as well as the honey they produced were utterly different. In 1880, the movable frame

FIGURE 1.1

Apis dorsata: the bee of the East. This particular bee was caught by US apiarist Frank Benton (or more likely by his local assistants) in India in 1904 during his global travels to find the "ultimate race of bees." This journey included the examination of honeybees and queens in Palestine, and their export to the United States and Java. *Source:* James P. Strange, "A Severe Stinging and Much Fatigue—Frank Benton and His 1881 Search for Apis dorsata," *American Entomologist* 47, no. 2 (2001): 112–116.

beehive—hailing from North America—arrived in Palestine. Replacing the local fixed clay hive, the success of this technology was sweeping. Bees soon had new mobility, with some bee queens traveling to North America and others reaching Java.[3] While bees continued to fly, collect nectar and pollen, and make honey that humans enjoyed, their labor and behavior dramatically changed. Furthermore, although this technology gained prominence in many other parts of the world, the movable frame beehive in Palestine had a unique role as it served as a tool for the production of scientific truths and reconfirmation of religious beliefs.

This chapter uses the story of the Baldenspergers, a family of settling Christian missionaries that were the first to use this hive in Palestine, to demonstrate how the movable frame beehive was transformative in late Ottoman Palestine, as it helped show that the land was literally flowing with honey. As part of the work of experts, this technology supplied evidence to prove that the land of Palestine was in fact unusual, becoming a Holy Land again, and that technoscience played a critical role in the process. Through the changes in beekeeping practices, I examine how the combination of mystical and scientific ideas was at the basis of large-scale changes in and domination over the land.

A BEE IN A BONNET

In the last decades of the nineteenth century, bees became the center of attention in attempts to re-create a land flowing with milk and honey for one family of German French missionaries, the Baldenspergers, who settled in Artas (or Urtas), a small Palestinian Arab village near Bethlehem. While the family began by practicing Indigenous beekeeping in Artas, it soon adopted a new type of hive: the movable frame hive. The family then started moving bees from Artas throughout Palestine, introducing these bees to new plants and ultimately transforming honey production in the area.

Caroline and Henri (Heinrich) Baldensperger (1823–1896) came to Palestine from Alsace through Basel in 1848 in order to join the German Swiss St. Chrischona Pilgrim Mission.[4] This mission was seen as the first step in "the establishment of Christian settlement that would be an example and a source of light to their surroundings." It brought the Baldenspergers, along with a few other missionaries, to Jerusalem as "craftsmen and peasants to the Holy Land, as salt to the earth and a light in the darkness." The missionaries, newcomers to Palestine, were to teach local populations practical knowledge, as the formal goal of the project was "not to send preachers to Palestine but people who would demonstrate true Christianity to the inhabitants through quiet work and good deeds."[5] Together with a locksmith and mechanic, soap maker and chemist, and a watchmaker, Henri was to demonstrate true Christianity in Palestine.

A few months after the couple's arrival to Jerusalem, however, Henri left the mission and moved to the Bethlehem area, where he lived until the head of the mission asked him to return.[6] According to his son, Philip, Henri and Caroline operated under "the belief that they were called, under the protection of Divine providence, to teach the people of Palestine better ways, not by preaching the Word, but by exemplary life and work." The couple built a house in Artas, an Arab village inhabited by approximately two hundred people, to initiate their independent mission "among the natives."[7] The Baldenspergers had eight children—Théophile (1851–1939), Charles Henri (1853–1905), Philip J. (1856–1948), Emile (1858–1946), Jean (1860–1911), Louise (1862–1938), Willie (1865–1891), and one who did not survive; they grew up in Jerusalem and the village, among the villagers and occasional European visitors. In old ruins around this village, and with the guidance of the local "Master of Apiculture"—a Palestinian man named Jadallah—Henri also started keeping bees.[8]

Beekeeping in this region has a special role in the global history of bee-keeping given that the earliest evidence for organized honey hunting was found in Egypt. According to European sources, beekeeping methods and hives in the region, particularly in Palestine, resembled traditional Egyptian ones.[9] Although hives varied in shape (Palestinian hives were usually cylindrical), they were all fixed clay hives with back opening. Jean Baldensperger described the hives the family encountered, noting that "in almost every village of Palestine and Syria bees are kept, and, with a very few exceptions, they do not keep such numbers to depend upon them for their living, but simply a few hives placed one on top of the other, having arch built over them or some protection intended to keep away the hot sun-rays."[10]

Other sources hint at the practice of beekeeping in Palestine, especially in Artas. A 1924 ethnography of Louise Baldensperger's Palestinian maid, A'lia, for example, gives supplemental evidence for beekeeping in the village.[11] In her story, beehives are listed as part of the greater family fortune.[12] Furthermore, in an article published in 1888 in the *American Bee Journal*, Jean details the intricacies of Palestinian beekeeping methods while demonstrating a typical Orientalist disregard for the local practice:

> In general, bee-keeping is carried in very primitive and negligent ways in some respects, as weak colonies are never cared for. . . . The only work performed is in the swarming season, when swarms are watched for a few weeks in April and May, and hived into clay cylinders. The back covers are put on after hiving, and besmeared with wetted argillaceous earth. The interior is rubbed with citron leaves, and the small fly hole stopped with a few herbs for a day or two. They are then released, and not again looked to till the honey-crop. . . . [The] general honey harvest is the September crop. The covers are then hastily broken open. A few puffs of smoke from the pitcher-smoker . . . are blown on the bees, a comb or two of honey is cut out and put away, the cover is immediately replaced, and the bees are left for a whole year.[13]

FIGURE 1.2

Charles Henri Baldensperger and clay hives in Palestine. *Source:* Ludwig Armbruster's Bildersammlung zur Bienenkunde (Ludwig Armbruster's apiary photo collection), on permanent loan from the University of Hohenheim to the Domäne Dahlem in Berlin. This image also appears in Phillipe Marchenay, *L'Homme et l'Abeille* (Paris: Berger-Levrault, 1979), 59. I thank Falestin Naili for her help in deciphering who appears in this image.

This critical colonial view of beekeeping, which portrays local practices as careless and disorderly, nevertheless highlights the prevalence of beekeeping in the area and explains the limitations of the honey production. Given the fixed structure of the hive, honey production in Palestine was limited to local, seasonal swarming. In his first days as a beekeeper, Henri was trained in and used similar Palestinian methods. According to Philip, his father "kept bees . . . in the old castle above Solomon's Pools beyond Bethlehem, in the old clay hived of immemorial model" (see figure 1.2).[14]

In addition to keeping bees, most of the Baldenspergers—that is, Henri and his children—were occupied with writing about beekeeping and honey production in professional apiculture journals published in France, Britain and the United States. Family publications, however, addressed issues beyond beekeeping, and Philip's and Louise's contributions in particular are considered of great value to the study of Palestinian life during the late Ottoman period. With the great scarcity of surviving written materials from this time, the Baldenspergers' scope of work supplies a mirror, albeit settler colonial, into the Palestinian environment and culture. While Louise's 1932 book *From Cedar to Hyssop* focuses on documenting the plants of Palestine, the 1913 work of Philip, *The Immovable East*, for instance, concentrates on the Palestinian people with their customs and points of comparison to the scriptures.

With the opening of the Anglican missionary school on Mount Zion in Jerusalem in 1853, Henri was called back to service and became a housefather at the school for orphan Arab boys.[15] It was mostly his children (and European guests, such as Finnish anthropologist Hilma Granqvist) who inhabited the house in Artas throughout the years, worked the lands, and gradually became interested in apiculture. In their various publications, Henri's children explicitly state that beekeeping in Palestine had a special value, which depended on two kinds of justifications. First, identifying Palestine with the biblical land required continuous comparisons of the scriptures to the land and people of Palestine, and second, the product of the land had to become bountiful.

Throughout this period, in both kinds of their publications—professional apiary journals as well as their ethnographic and botanical work—the Baldenspergers put great effort into identifying parallels between the land they encountered in Palestine and the land of the Bible. Philip, the most prominent and prolific writer in the family, was a frequent contributor to the *Palestine Exploration Fund Quarterly Statement* between 1890 and 1920, thus participating in British efforts to survey the topography and ethnography of Palestine in relation to the biblical land during the turn of the century.[16] Prior to focusing on the practices of Palestinian people, Philip concentrated on demonstrating specific connections between Palestine and the biblical land, as in his article titled "The Identification of [Biblical] Ain-Rimmon with Ain-Artas (Urtas)."[17] Jean's 1888 article, "Palestine: An Account of Bee-Keeping There by Eye-Witness," which was published in both the *British Bee Journal* and *American Bee Journal*, is similarly saturated with comparisons between different places and practices in Palestine and various references in the scriptures.[18]

Granqvist, who stayed with the Baldenspergers in Artas while researching the people of the village, wrote about the "biblical danger" of comparing Palestine to the biblical land. In her discussion, she clearly criticized the most famous of Philip's works, arguing that "[there is] temptation to identify without criticism customs and habits and views of life of the present day with those of the Bible, especially the Old Testament. Only too often one has been tempted to build a bridge from the past to the present by combining modern parallels with Bible versus . . . a period of 2000 years and more between them—a gap which cannot be explained away merely by citing 'the immovable east.'"[19] Not only is this critique telling of Philip's agenda and the way his writing was perceived, but it is a strong indication that the so-called biblical danger was a prevalent phenomenon among European travelers to and settlers in Palestine.

Beyond comparing locations in and habits common to Palestine with the scriptures, the identification of Palestine as the Holy Land depended on providing evidence of its fertility. If Palestine was the biblical land, could it also flow with honey? In their writing to professional apiary journals, Jean and Philip frequently compared references to honey in the scriptures to their findings in Palestine, and analyzed the inadequacies between the scriptures and reality. They argued that specific changes in beekeeping allowed for the abundance of honey in Palestine and highlighted the family's role in enabling these changes. The Baldenspergers indeed adopted a transformative form of beekeeping. Their story, however, is but one of a larger movement. According to the common understanding of the time, Palestine was the sacred land of the Bible and as such it held special qualities. For this land to become sacred again, particular practices—rational, modern, and scientific—had to be employed by particular people. Technology and new forms of movement were interlaced with religious sentiments.

REDEMPTION TECHNOLOGIES

Technology was central to the age of colonialism, often acting as its ideology.[20] It is commonly assumed that with the Industrial Revolution, science and technology replaced the traditional role of religion as the central justification for conquering new lands.[21] In fact, European settlers in Palestine, both Christians and Jews, frequently acknowledged their scientific and technological advantage over Palestinian people. Yet religious motivations continued to play a crucial and explicit role; it was the combination of religion and technology, and the pairing of past notions and ideas about future developments, that grounded their efforts to create a land of plenty, not least because early settlers were religious people whose work was framed by a mission. Put differently, if technology acted as the ideology

of colonialism, missionaries were its common agents of execution. As the
work of colonial history scholars demonstrates, missionaries were key to
fashioning structures of power and inequality—a symbolic and material
process that evolved through the everyday world of the colonial encounter.[22] For missionaries and other settlers in Palestine, the daily path to
redemption was paved by tilling the land. A description of early European
settlement efforts in Artas illustrates the spirit of these settlers and the perceived importance of their practices:

> The débris and rocks of former terraces ten miles east of Jerusalem, while they ren-
> der cultivation under the present method out of the question, were at the same
> time the downfallen monuments of the former industry and prosperity of the
> people. But the efforts . . . of "the industrial settlement" near the pools of Solo-
> mon, southeast of Bethlehem, enable us to add to the above the facts of present
> produce. . . . This may explain the wonderful fertility predicated of this country
> by early writers, and which seems to be so poorly sustained by the appearance of
> the land at the present day.[23]

From the early years of European settlement in Palestine, great efforts
were put into increasing the production of the land, but the choices behind
such products were highly calculated. Honey was such a desirable product,
the beehive the chosen technological means, and the honeybee the center
of attention and manipulation. In their writing to the Western beekeeping community, the brothers glorify the Palestinian bee, which they often
name "the holy bee": "The Palestine bees are good honey-gatherers, and
the queens are very prolific and beautiful. It is not rare to have colonies
yielding upwards of 100 pounds for a single crop, through it is not the
average."[24] While the Baldenspergers frequently acknowledged the value
of honey produced in Palestine, they also lamented the technical limitations inherent to the system. In her book *From Cedar to Hyssop*, Louise discussed Indigenous beekeeping methods, which she perceived as simple and
primitive:

> Village beekeeping in Palestine is a constant source of surprise and interest to
> newcomers in the country. The bees are housed in clay pipes built up into stacks,
> placed usually inside the village in courtyards or on low roofs, and they often
> have a very picturesque appearance. They are like the beehives of Egypt, which go
> back . . . to at least 2600 B.C. . . . The honey . . . is often of excellent quality, but
> marred by dirt owing to primitive methods of dealing with the comb. . . . In spite
> of these primitive hives, bees seem able to live healthily in them; at least there is
> no record of disease before the introduction of infected bees from South Russia.

But the system has one defect: the hives are fixed, and the crop is therefore limited to what the bees can get near the village.[25]

Although Louise explained the virtues of the Palestinian honey, she pointed to its imperfections too, highlighting the immobility of the Palestinian hives. Thus in spite of the natural qualities of the bees and land, beekeepers in Palestine faced constraints. As the Baldenspergers came to exemplify, providing the proofs for a biblical land as well as the supremacy of its crop were not sufficient. Instead, in their view, particular European interventions were necessary in order to achieve the ultimate productivity of bees and an abundance of honey.

According to family records, a crucial change occurred in the late nineteenth century. In 1880, Henri met two prominent North-American beekeepers exploring the area, D. A. Jones and Frank Benton. Given the recent success in importing the Italian honeybee to North America, the late nineteenth century was a period of intensified efforts to identify new species of bees. Jones and Benton were part of this growing entrepreneurial trend, and they traveled throughout Europe and the Middle East with the intention of learning, breeding, and eventually exporting bees (and queens specifically) to North America.[26] The two introduced their findings to the beekeeping community in the United States and made special efforts to distinguish the holy bee from other regional species.[27] In the midst of this attempt to find the ultimate bee, the Baldenspergers were introduced to a new technology: the movable frame hive.

Global beekeeping went through a series of fundamental changes during the nineteenth century. Following concurrent developments in Europe and the United States during the seventeenth and eighteenth centuries, a specific model of hive known as the Langstroth movable frame hive received great recognition and was gradually adopted globally in the later nineteenth century.[28] The structure—a square wooden hive containing several wooden frames in which bees could build a honeycomb—assured great advantage over the existing local hives. In the Langstroth model, the wooden frames were easily removed and placed within the wooden boxes, which were positioned on top of each other. The structure forced the bees to build the comb on the frames alone, leaving the box and gaps between the frames detached. This allowed the beekeeper to remove the frame from the box and extract the honey without breaking the hive altogether, as was done in various hives globally. Avoiding the destruction of the hive and

FIGURE 1.3

The Baldenspergers demonstrating the use of the Langstroth movable frame beehive alongside local clay hives, 1890. *Source:* Armbruster's Bildersammlung zur Bienen-kunde. Reverend L. L. Langstroth (1810–1895) patented his model in 1852 and wrote two decades later, "If every bee-keeper would adopt this plan [of hive], our country might soon be like ancient Palestine, 'a land flowing with milk and honey.'" L. L. Langstroth, *The Hive and the Honeybee: The Classic Beekeeper's Manual* (Mineola, NY: Dover Publications, 1878), 182.

combs not only increased the beehive's life span but also directed bees' energy into building new combs to produce more honey. These changes were therefore central to the success of the new model, as it allowed for a great increase in the cultivation of honey.

Thus not only was each frame movable but also so was the entirety of the beehive.[29] Pastoral beekeeping now synced the transport of bees throughout Palestine with the blooming of plants so that the bees could make honey yearlong. The new movement of bees was part of an emerging large-scale technological system that included honey-making machines, but irrigation pipes, roads and cars, and different kinds of workers too.[30] The US explorers left the region shortly, and the community of global bee-keepers promoted the Italian bee as the best of breeds. But the Western way of keeping bees gained prominence in Palestine: long-distance bee travels became commonplace in Palestine, and the honey products—with flavors now including thyme, prickly pear, and citrus fruits—reached European markets.

Ultimately the combination of the movable frame and movable hive was of great influence on honey production in Palestine. In her 1932 book, Louise described how the new technology transformed her family's method of beekeeping: "Nowadays modern beekeeping is spreading in Palestine, in the Jewish colonies, and among the Arabs too. It all started by the Baldensperger brothers with their introduction of the first movable hives in 1880. They were the first to have the brilliant idea of carrying them about, from coast to the hills, and so assuring a crop of honey all through the season."[31] The success of the movable frame beehive was sweeping; to this day, beekeepers throughout the world use this nineteenth-century technology. In Palestine, though, at least for a short while, and prior to the widespread construction of roads and availability of cars, bees lived inside the new movable hives but traveled on the old transport technologies (see figure 1.4).[32]

While Palestine's fixed hive fit the Western image of Palestine as "the immovable East," as described by Philip Baldensperger, movement existed all along in various, often forgotten ways. The shift from static to movable hives depended on prevalent forms of movement. Figure 1.4, which shows bees traveling on other animals illustrates how—if just for a brief moment in time—the new movement of bees was made possible by the old movement of camels; it captures what historian David Edgerton calls "creole technology," showing how new technologies of movement were combined with existing and established ones.[33] Moreover, according to one account, movable beehives were carried on the heads of women across the country, one hive at a time.[34]

Together with the anonymous figures on both sides, camels and local women were essential for the eruption of new movement. Beyond the new technology, then, this image reveals some of the actors who remained invisible forces for change in the East. Hence new technologies, such as the wooden frame hives, were successful in Palestine not only for the reasons behind their global acceptance but also because they became intertwined with old, local (and often living) ones. Furthermore, long after camels were deemed unnecessary for (or even interfering with) moving bees, backstagers continue to be locomotors; as such, in this sense, all technologies are so-called creole technologies.[35]

FIGURE 1.4

Bees on a camel on their way to Jaffa in 1890. *Source:* Armbruster's Bildersammlung zur Bienenkunde.

CHANGES IN THE HIVE

Louise herself highlighted some of these new-old combinations: "When the time came to take the bees from the Orange blossom of Jaffa to the thymy uplands they bound the hives on a camel and proposed to travel by night while the bees are asleep."[36] The new mobility of the hive meant that beekeepers had to take the transport of both camels and bees into consideration. For example, in order to move the beehives around, extend the swarming season, and ultimately produce more honey, pastoral beekeeping had to develop new strategies—such as traveling on camels at night while the bees were asleep—for managing the population of bees and their temper. The Baldenspergers shared amusing stories about frightened camels and alarmed employees, adding, "Nowadays bees travel swiftly by motor lorry and the excitements of other days are no more, or perhaps we might say changes, for we think that beekeeping is never a very quiet kind of occupation."[37]

For the Baldenspergers, however, beekeeping became more than an entertaining hobby. Philip described how meeting Jones and Benton changed the Baldensperger brothers' mind about beekeeping completely. In addition to adopting the movable frame hive, this new form of beekeeping required training. Shortly after the meeting, Philip joined Benton's "bee-conversion" in Beirut and Cyprus, where "the most important work done . . . was the adoption of a standard frame" and "modern methods in beekeeping" had taken permanent root.[38] After Philip stayed for "many months and thoroughly learned apiculture," the brothers finally decided to "abandon their agricultural work, let out the family lands on hire, and devoted themselves exclusively to bee-keeping."[39] Thus following Philip's professional training, the Baldensperger brothers adopted beekeeping as a main occupation, although each did so in different ways. Granqvist noted that Henri "hoped that his sons would live in the village . . . but neither did anything come to that. Two of his sons Phillippe and Emile Baldensperger who had learnt bee-keeping were in Artas for a short time but then settled in other places; Phillippe in Nice and Emile in Jaffa where they continued with their bees."[40] While Louise remained at home in Artas, modern beekeeping took the bees as well as the Baldensperger brothers elsewhere.

"Pastoral beekeeping" involved other struggles beyond that of transportation. Philip portrayed the realities of disease, taxation, and government intervention—using the kind of imagery that would haunt the memory of the Ottoman rule in Palestine—explaining that they had to fight "mosquitoes and the fever—a consequence of roaming about in unhealthy marshy places—as well as the vile tax-gatherers and Turkish officials' odious vexations and injustice."[41] In addition to the tragic death of their young brother, Willie, who drowned in the sea near Jaffa in 1891, health problems and the supposed struggles with the Ottoman authorities finally led most of the brothers to leave Palestine and continue their work in North Africa and Europe. Emile and Jean "carried part of their hives and apparatus with them to Algeria," where beekeeping proved more profitable, and Philip "exhausted by fever and doubtful of ever being able to change the mentality of the natives," moved with his family in 1892 to Nice, France, where he established an honorable career as a beekeeper, researcher, writer, and lecturer.[42] Named in old age as "Père Baldens," Philip became a renowned scholar and contributor to the global knowledge of bees and beekeeping.[43] With the death of Philip's daughter, Nora, herself an esteemed apiarist, in 1977, the British *Bee World* Journal commented that her father was "one of the greatest international figures in beekeeping between the wars."[44]

In *The Immovable East*, Philip mentioned that Emile and Jean, who moved to Algeria, "were soon glad to return home [to Palestine] again, as it is still 'the land flowing with milk and honey.'"[45] Emile was the son to continue on with beekeeping, and established a successful independent apiary in the Jaffa area, the main economic artery of the region, and continued to move his hives from place to place according to seasonal flowering (see figure 1.5).[46]

Emile's beekeeping practice not only enabled his "intimate contact with the people in very different parts of the country" but also exposed the bees to the blossoming citrus trees of Jaffa at a period when the citrus industry was booming and European settlers were increasingly dominating it.[47] This strengthening relationship—between the bees and citrus trees—in the process of seizing European control over the citrus industry, brought the bees and oranges much acclaim.

ABUNDANCE AND ITS DISCONTENTS

In her analysis of the plants of Palestine in *From Cedar to Hyssop*, Louise wrote about plants that became "important for the bees" in the new form

FIGURE 1.5
Emile Baldensperger, a professional beekeeper with his movable frame hives in the Jerusalem area, photographed by his sister, Louise. *Source:* Grace M. Crowfoot and Louise Baldensperger, *From Cedar to Hyssop: A Study in the Folklore of Plants in Palestine* (London: Sheldon Press, 1932), plate 9.

of beekeeping. She highlighted that "of all these crops, the Orange Blossom is the most important. Mr. E. Baldensperger tells us that once in an exceptional year—1883—ten hives gave a total of a little over 3,000 pounds of honey in Jaffa!"[48] Other sources acknowledge the great success of the family practice, even in comparison to global standards. In 1884, Henry Allen, a famous US beekeeper and researcher, wrote in the *Bee Journal* that "with only 50 and 60 colonies they [the Baldensperger brothers] had taken 5,800 pounds of honey, 5,200 of which were taken in 16 day. How many apiaries with this same number of colonies in America can make such a good report?"[49]

Maintaining and expanding the scope of the new form of honey production depended not only on the type of bees or variance of plants but the regulating authority as well. In his memoir of traveling in Palestine, British literary scholar and Qur'an translator Marmaduke Pickthall mentioned "a

French Alsatian family, the Baldenspergers, renowned as pioneers of scientific bee-keeping in Palestine." Pickthall described the Ottoman attitude
toward Western beekeeping in Palestine and way this attitude changed:

> They [the Baldenspergers] had innumerable hives in different parts of the coun-
> try. . . . For a long while the Government ignored their industry, until the rumour
> grew that it was very profitable. Then a high tax was imposed. The Baldenspergers
> would not pay it. They said the Government might take the hives if it desired to
> do so. Soldiers were sent to carry out the seizure. But the bee-keepers had taken
> out the bottom of each hive, and when the soldiers lifted them, out swarmed the
> angry bees. The soldiers fled; and after that experience the Government agreed
> to compromise.[50]

Philip, too, referred to this change when he noted how Emile's beekeeping generated more success along the years "as the Turkish officials have
become more accommodating."[51] In spite of the initial objections, according to such testimonies, Ottoman authorities gradually tolerated such
changing form of beekeeping.

Since the new form of beekeeping demanded considerable financial
investment—partially because movable frame hives and other equipment
were usually imported from Europe or the United States as opposed to
the domestic, handmade clay hive—Ottoman support for this new form,
together with the technical benefits of the movable frame hive, established
an advantage for European ways of honey production.[52] As early as the turn
of the century, therefore, the project of creating abundance in Palestine was
beyond the hands of Indigenous people and to a great extent adopted by
the changing governing powers.

The new mobility of the hive, however, also posed a threat to European ways of beekeeping. As beehives became movable, and as honey
production—now monopolized by Europeans—grew accordingly, reports
of theft of bees and hives grew as well. The *Haskell Free Press* in Texas, for
example, published "Famous for Honey: An Industry of Palestine in Biblical
Days May Be Revived" in 1900. This article mentioned the Baldenspergers'
work and discussed the difficulties that accompanied their success:

> The greatest enemy [to beekeeping in Palestine] is man. Whenever an apiary is set
> down, the sheiks of the nearest village have to receive a certain amount of honey,
> otherwise the bees will be stolen. . . . [A]bout one-tenth of the honey produced
> must be given away to prevent people from taking the hives. Furthermore, when
> the bees are carried from one place to another on camels, the Bedouins, or wild
> Arabs, occasionally steal the camels.[53]

With the growth of new beekeeping practices, theft became common-place throughout Palestine. Numerous beekeepers complained about the disappearance of their hives over the years, connected stealing to the local struggle against European settlement and the Zionist movement, and argued that the different governments handled the matter inadequately.[54] Mobility became a two-edged sword for some; thanks to the new trans-portation of the hive, stealing bees and honey-making machines allowed Palestinian Arabs to resist—as well as configure—this process of change in the land and the new means of owning it.

PROPER, SUPERIOR HIVES

From the early years of British rule in Palestine, the government officially supported the expansion of honey by the complete removal of taxation over "bee-hives, hive frames, honey extractors, centrifugal machines for honey extraction and hive foundation."[55] The British government also offered loans to those who wanted to become beekeepers: "The loans" it was declared, "will not be given in the form of money but in the form of hives, equipment, and wax."[56] Moreover, throughout the 1930s, the British government supplied sugar tax free or discounted for the sake of feeding bees. The support for beekeeping in Palestine, however, depended on the use of the movable frame beehive: "Bee-keepers are hereby notified that Government [sic] has decided to sell duty-free sugar to those bee-keepers who are in possession of populated movable frame comb hives."[57]

Under British rule, as part of agricultural development programs, bee-keeping was explicitly promoted, professionalized, and standardized.[58] Starting in the late 1920s, the government established professional training courses at the governmental agricultural station in Acre.[59] Jewish agricultur-alist Alexander Livshutz, who had previously trained with the Baldensperger brothers, organized the training courses, which demonstrated the benefits of the movable frame beehive and delegitimized the static clay hive:[60]

> Modern methods are essential, and the Government Department will be glad, at its various beekeeping stations, and especially at [the governmental experimental farm in] Acre, to give all enquirers the fullest information and advice on the sub-ject, to sell bees of good stock . . . and in particular to give practical demonstra-tions of the superiority of movable comb-hives over mud-hives. With mud-hives, the bees have often to be killed before the honey can be extracted, and then it

is full of impurities; with movable comb-hives, the honey can be removed in its purity without harming the comb or the brood, and natural swarming can be controlled and the number of colonies of bees increased as the beekeeper wishes.[61]

Along the lines of the colonial logic, the new hive was equated with modernity and mobility was equated with progress.[62] The promotion of a different temporal standard through the use of movable hives, or the attempt to "expedite the peasants," as one article called it, was well organized: "Last Friday the Mukhtars [appointed heads of villages] of the Jerusalem district were gathered for a meeting with District Officer Nicola Saba. He demanded that they start using new methods in agriculture, and become interested in raising chickens, bees, etc. The government is willing to help the *fellaheen* [Arab peasants] in purchasing proper hives instead of the clay vessels."[63] The contrast between the mobility of the new hive and immovability of the old, in the eye of officials, mapped onto and naturalized the contrast between the European settlers and native Palestinians. A US report, sent by the consul in Jerusalem to the US State Department in 1921, noted that "an attempt was made to interest the natives in these modern methods but without any appreciable success. . . . In later years, however, the Jewish immigrants to Palestine who came here to found agricultural colonies have taken up honey production on modern lines." The successful creation of movement was not only a test for the tools of European modernity, though. According to this same report, the promotion of European beekeeping became a way to examine if Western technologies could fulfill the potential for a plentiful land—a land flowing with milk and honey—through their literal flow.[64]

Official support for the production of honey in Palestine signified a great step in the making of a fertile land—one that went beyond the hands of Palestinians and one missionary family that appreciated honeybees. The Baldenspergers continued to practice this new form of beekeeping in Palestine and elsewhere, and participated in the global apiary community. Yet with time, and in comparison to earlier years, they gradually came to doubt and finally reject the connections between the biblical metaphor and its manifestations in Palestine. For example, the earlier writings of the Baldenspergers insisted on making connections between metaphor and reality in the land of Palestine; in his 1888 article, Jean claimed that "honey was always abundant in Palestine."[65] Over the years, however, the family expressed different opinions in regard to the land's abundance. In 1932,

Louise contended that her brother Philip objected to the idea that biblical Palestine was flowing with honey, noting that "surely we may still speak of Palestine as a land flowing with milk and honey, even if, as Mr. P. Baldensperger declares, 15 out of 19 references to honey in the Bible are more likely to mean dibs, grape treacle. . . . He goes far as to say that in his opinion bees were not brought in till after the Captivity."[66]

In an article published in 1931 in *Bee World*, Philip deepened his doubt in the biblical ancestry of bees and honey, citing that "honey was not known to early Hebrews, nor were bees." As opposed to others that "accept . . . 'Canaan as a land flowing with milk and honey' as literally correct," he explained his change in approach at great length:

> I must say also that from boyhood I was amongst the fervent Bible readers on Mount Zion, and still continue to be so, I admitted the "Land flowing with milk and honey" as Oriental mentality puts it, to be very nearly correct. This becomes very different when we look at it not poetically but practically and as a beekeeper; and I therefore studied every nook and crevice, every local name . . . and ask "could honey possibly flow down the face of rocks without calling forth the most tremendous robberies by bees ever heard of?" Bees are sometimes crevices, but I never succeeded in Palestine to bring them forth.[67]

Throughout the years, therefore, the Baldenspergers expressed varied opinions in regard to the connection between the land of the Bible and modern Palestine. Their publications nevertheless demonstrate a wide and consistent interest in these connections, and the kinds of efforts that were put into making a metaphor a reality. By the time the Baldenspergers—who had established such parallels by way of practice—abandoned this effort, other settlers adopted it fully. As Philip told the Arabic readers of the Jaffa-based *Filastin* in 1921, "Jewish settlers, who attempted to revive the land they considered to be of their forebearers, became keen on beekeeping, and demonstrated a great interest in expanding the use of the movable-frame beehive in Palestine."[68]

Israel Robert Blum (~1898–1979), a Czech Jew who immigrated to Palestine in 1924, and became a successful apiarist and leading figure in the Jewish beekeeper community, was investing efforts in establishing similar connections between beekeeping and the biblical land. Around the same time that Philip was denying these relationships, Blum observed,

> It is not a poetic phrase that our land was called in Biblical times "a land flowing with milk and honey." My mentor and teacher—Prof. [Ludwig] Armbruster, the

famous bee researcher from Germany—that visited the land of Israel a year ago, wrote about this matter an essay called "The Bible and The Bee." He thinks that we should interpret the word "honey," which appears so often in the Bible, as bee honey and not as date honey. According to him, beekeeping was very developed and widespread in this land already thousands of years ago.[69]

Throughout his life, Blum continued to insist that biblical Hebrews were beekeepers. In 1951, for example, in *The Man and the Bee*, a guidebook that became a classic, he considered this belief a moral obligation:

> I have no doubt that the Israelites made sweets from dates, figs and other fruits that contain sugars. And yet additionally they were occupied with beekeeping so they have real honey. As an expert that is dealing with bees for more than fifty years and that devotes his mind every day to his work, and who has learned that our fathers when they lived on their country and in their land had incomparable knowledge in comparison to other people in the region, I must say that every person who sees the aromatic and pure honey that our father collected from the honeycomb an artificial product, is not respecting the memory of our fathers as should be.

Jews were not alone in making this argument, he posited: "It is worth mentioning that those most educated in the occupation of bees among the gentiles . . . see the honey as real honey as it says in the Bible."[70]

The emphasis on such connections was not only limited to Blum and his gentile mentor but also widely accepted among the growing Jewish professional beekeepers' community in Palestine. The hall of the 1939 Jewish Beekeepers Conference, for instance, was decorated with "verses from the Bible, which detail the value of honey."[71] Moreover, Friedrich Simon Bodenheimer (1897–1959), a famous Jewish German entomologist and zoologist, and later a historian of science, gave an opening lecture in a professional course for Jewish beekeepers in 1933 titled Beekeeping among Jews in Biblical Times.[72]

Professional conferences became recurrent sites of debate over the probability of the existence of honey in biblical times. These debates were crucial, as the proof for the existence of beekeeping practices was considered a justification for contemporary Jewish beekeeping and more broadly the growing Jewish settlement in Palestine. In 1956 (after the establishment of the State of Israel), on returning from a meeting of the International Beekeepers Association, Blum offered his own insights into the Baldensperger rejection of the parallels between honeybees and the Holy Land:

This is the first time that a Jewish Israeli actively participates in the congress. In previous meetings there was always the spokesman of the Holy Land, Mr. Philip Baldensperger, a son of a missionary family, who was born a hundred years ago on Mount Zion. Him and his four brothers were the first beekeepers in our renewed land. . . . Mr. Baldensperger, that lived in Nice in recent years, was a respected figure among beekeepers around the world. There is no wonder, therefore, that he managed to promote the idea among them that in biblical time our fathers had no contact with bees. His remarks regarding our current ways of beekeeping [in Israel] were also very malicious. The basis of this perception he holds becomes clear if we remember that his brother Emile, that lived in Ajami [an Arab neighborhood] in Jaffa, was the father-in-law of Mr. Rock, who was a member of the Supreme Muslim Council, of the Mufti Haj Amin al-Husseini.[73] Clearly, in my lecture I saw it as my primary duty to contradict Mr. Baldensperger's popular argument and reestablish the honor of our beekeeper-fathers as well as the beekeepers of our generation. Using many biblical phrases . . . accompanied by drawings, I proved that not only did our fathers keep bees, but also that they had a lot of professional knowledge in those days.[74]

Thus Blum cited personal ties with Palestinian Arab nationalists—seen as evidence of support for the Palestinian cause—as the reason behind the Baldensperger rejection. According to Blum, redemption was now the duty of Jewish Israeli beekeepers. Philip, on his end, argued that settlers' enthusiasm for and success in beekeeping depended on European funds and support, without which "Palestine will be for Palestinians."[75]

SWEET DISORDER

The story of the Baldensperger family at the turn of the twentieth century is an example of the ways in which a biblical notion became an inspirational and legitimizing force for environmental, economic, and ultimately, political change. Here we see how the adoption of a new technology resulted in a sharp increase in the production of honey and a change in domination over the output of the land. Although the new practice of beekeeping was not unique to Europeans in Palestine, the combination of the technology and metaphor became a crucial component in justifying intensified European settlement in Palestine. Since a Western technology had seemingly succeeded in creating even greater flow of honey than the biblical one, European interventions were encouraged and felt justified. Europeans had succeeded in accelerating movement in "the immovable East."

The creation of biblical plenty went beyond honey. The history of milk production is a crucial part of the story (see chapters 2–4). Important too is the difference between milk and honey production; while the milk industry is considered a story of great success, beekeepers in Israel have increasingly lamented the deterioration of the local honey industry. They argue that the reduction in the production of honey was related to greater changes in the Israeli agricultural scheme. Just as the success of honey production was intertwined with the expansion of the citrus industry in the early twentieth century, the weakening of the citrus industry and gradual reduction of orange groves beginning in the midcentury (vis-à-vis an expansion in pine tree planting; see chapter 2) caused a decrease in the availability of nectar and hence a dramatic fallout in the production of honey.[76] Combined with expedited urbanization and recent anxiety for what's called the "colony collapse disorder," a sense of crisis in the beekeeping industry has emerged.[77] Another aspect of the analysis is the preference of milk and honey production over industries that were largely established and profitable in Palestine, yet lacked sufficient cultural currency. While particular industries were aggrandized as a result of European intervention, various salient industries weakened or diminished, as in the case of sesame seed and sesame oil production.[78]

Furthermore, the debate regarding the existence of biblical bees and their honey continues to occupy researchers today. Excavations in Tel Rehov from 2005 to 2007, for example, point to the existence of beekeeping in biblical times. An article in *Science News* noted in 2008 that "the Bible refers to ancient Israel as the 'land flowing with milk and honey,' so it's fitting that one of its towns milked honey for all it was worth. Scientists have unearthed the remains of a beekeeping operation at a nearly 3,000-year-old Israeli site, dating to the time of King David and King Solomon."[79] "It is a land of honey," argued the directors of the dig, Amihai Mazar and Nava Panitz-Cohen in describing how sophisticated scientific methods helped establish such a fact:

> The discovery of the beehives at Tel Rehov is unique since it appears that no apiaries have ever been discovered in the archaeology of the Old World. It comprises an innovation in the archaeological study of ancient economies in Israel and its neighbors during the period of the Israelite Monarchy. Based on the ceramic evidence and C14 dates, the apiary at Tel Rehov was in use during the latter part of the United Monarchy and/or during the initial period of the Northern Kingdom of Israel, prior to the Omride Dynasty.

The experts then explain the significance of their findings to the understanding of the Bible and ancient Hebrews:

> The term "honey" appears fifty-five times in the Bible, sixteen of which as part of the metaphor of Israel as "the land of milk and honey." This honey has been always understood as having been produced from fruits, such as dates and figs, with bees' honey mentioned explicitly only twice, both times in relation to wild bees. . . . However, careful reading of biblical metaphors mentioning honey led Forti (2006) to suggest that they refer mostly to bee's honey, through in her view, due to the lack of agriculture in the Bible, the references are to honey collected in nature. Indeed, in no case does the Bible mention bee rearing as a productive industry. The discovery of the beehives at Tel Rehov shows that this was a well developed economic branch during the First Temple period. We can now assume that at least some of the notations of honey in the Bible pertain to bees' honey.[80]

Acceptance of particular religious ideas and notions of past plenty continues to frame scientific work today; technological means, in turn, help in reifying these ideas.

2 GETTING THEIR GOAT

"We were informed that the government decided to eliminate the goats within five months [and] about the ban of herding goats in groves and forests," wrote the leaders of the A'ara village in a 1952 petition to David Ben-Gurion, the first prime minister of Israel.

> We do not wish to argue with the government in a matter in which it has a firm decision. Yet we want to draw your attention to the neglect and the harm that this matter entails for the owners of goats. . . . God created the goats for benefit and not for annihilation . . . and this governmental decision is against the will of god no more and no less. The government argues that the goats are the sworn enemies of the forests and the trees, but they never were.[1]

A handwritten petition, this statement was one of many sent in the 1950s to Israeli governmental officials by Arab goat owners, now under Israeli military rule (see figure 2.1).[2]

Goats were a central component of the Palestinian subsistence economy until the mid-twentieth century, and for many Palestinian villagers and Bedouins, their raising was an important occupation. Goats were the greatest population of domesticated animals in Palestine and had been the main producers of milk for hundreds of years in the area.[3] Governmental restrictions on goat raising and grazing in the early 1950s caused much anxiety and frustration for those who depended on these animals for their livelihood. For the few Palestinians who managed to stay on their land after the 1948 war, their lives changed dramatically; their property shrunk, and their practices and movements were controlled and restricted under the new regime. The government allocated lands previously used for grazing

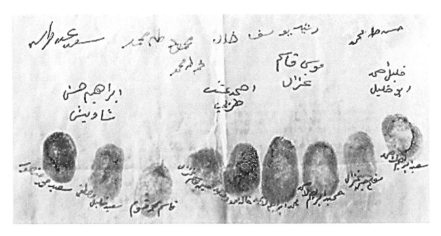

FIGURE 2.1
Thumb signatures of goat owners of Sakhnin village, as part of their petition to the
prime minister of Israel, December 28, 1952. *Source:* Israel State Archives [ISA]/Gimel
Lamed-19/17022.

animals and crop planting as either closed forest areas or Jewish agricultural
settlements. Moreover, it attempted to systematically reduce the number of
goats in the country in order to minimize their perceived environmental
damage by issuing the Black Goat Law in 1950. Some Palestinian villagers
from A'ara and elsewhere within the borders of the State of Israel chose to
petition to Israeli government officials and other state institutions to resist
this changing reality.[4] They argued against the limitations on movement
and the usage of space, rejecting the claim that these were necessary for the
sake of reviving the land.

This chapter deals with the gradual process by which the local milk producer, the goat, came to be seen as *the* enemy of nature, a hindrance to
the revival of Palestine, and a threat to the social order by both British and
Israeli rules.[5] I examine the negotiations over the meaning and habits of
this animal, which was long a symbol of the diasporic Jew and then gradually became a symbol of the rebellious Palestinian Arab peasant. In essence,
this chapter deals with the vilification of the local goat and its owner, and
how within the framework of the law, Palestinian villagers and other historical actors objected to this process and its basic theoretical assumptions,
and challenged the desire to transform the land and its creatures. This is a
story of how the best candidate for producing plenty of milk in Palestine/
Israel became an outcast.

Recent studies of the Middle East/North Africa region use the tools of science and technology studies and environmental history to talk about colonialism and the image of the East through Western lenses.[6] As part of this trend, some work has pointed to the tendency to believe that the Middle East became a desert as a result of the behavior of those native to the region. In a similar manner, as part of the attempts to revive a particular understanding of the Holy Land, natives to the land of Palestine/Israel were marginalized and blamed for its destruction. The chapter reveals the growing disparity between experts' ideas about the land and the lived experiences as well as violence and ironies entangled with this process of scapegoating.

DEVOURING CREATURES OF EMPIRE

Limitations on grazing did not begin with the establishment of the State of Israel in 1948. While the Israeli rule exacerbated them, these restrictions were long known to shepherds and goat owners in the area, both Arabs and Jews. Grazing laws had developed rapidly during the British Mandate in Palestine after World War I. As British rule stabilized, officials implemented their policy of afforestation, which was considered to be a pivotal aspect of improvement and development, and a major way to utilize land resources throughout the entirety of the British Empire.[7] When World War I concluded, the British thus established the Department of Forestry in Palestine, which was responsible for fostering its forests and, more broadly, taking care of nature.

Britons regarded the planting of forests as a crucial step in developing the lands under their rule.[8] Furthermore, British ideas about afforestation and development went hand in hand with the growing planting fervor of Jewish settlements in Palestine, led by the Jewish National Fund (see figure 2.2).[9] While foresters certainly appreciated trees, there were other creatures that loved them even more, particularly herds of local goats, who showed great appetite for young pine trees. The hunger that goats demonstrated was so great that grazing came to be seen a threat to the maintenance and growth of forests, a cause of flooding and soil erosion, and hence a major hindrance on the way to development. "The complex processes of nature depend for their successful continuation on the system of balance," noted one British report from the early 1940s, using a common ecohistorical argument:

FIGURE 2.2
Jewish laborers planting forest trees during the British rule in Palestine. "Commencement of Afforestation Work in Guara Village, Communal Settlement," 1937. *Source:* Lazard's "Holy Land" Collection, Herbert Katz Center for Advanced Judaic Studies, University of Pennsylvania.

In the earliest times Mediterranean countries were covered with forests, but as soon as he [man] successfully emerged from his early stages of development, he began to upset the balance of nature. . . . [He] found it better to depend on flocks of goats. . . . It is a fact that the goats, camels, and sheep are the primary cause of flooding and soil erosion in this country. . . . [O]vergrazing prevents the beautification of the country with roadside trees, and the planting of shade trees in the villages. . . . [P]ractically all rural land . . . [is] patrolled by flocks of small agile goats which are ready to climb rocks, bushes and any obstacle in order to devour all green vegetation.[10]

Destructive and nasty, goats played a major role in this interruption of the balance of nature. To the British, control of that balance required control of Palestinian goats.

The old legal system was an anchor in handling the goat problem. While the British government encouraged the expansion of tree planting and discouraged grazing in many Mediterranean colonies, the Forest Ordinance in Palestine and the following environmental laws were based on old Ottoman rules. In addition to favoring the planting of trees, these laws reflected the government's main concern: local unrest and turmoil. Eager to avoid conflict, and in spite of British impressions of their own legal superiority, British officials continued with Ottoman rules. M. C. Alhassid, for example, noted, "It is important that when amending the Forest Ordinance it should be made clear that it is re-stating existing legilation [sic] which is contained in the Ottoman Land Code, and is not merely a new enactment which is not retroactive. Otherwise many difficulties may be anticipated."[11]

While the British government acknowledged the importance of goats and goat grazing to the local economy and people's livelihoods, and feared the resistance that would arise with growing limitations, the evils of the goat necessitated gradual legal change. After all, "the old land laws and customs were based on the common error that the goat is the friend of the poor man."[12] Faced with what they considered to be desolate and barren land, British officials believed that Palestine's revival depended on a systematic shift in the order of the land. F. R. Mason, acting chair of the British Soil Conservation Board, wrote in 1946 to all district commissioners that "the devastating damage done to this country by indiscriminate free range grazing cannot be stressed too often. It is hoped that all the State Domains allocated as forests will in time be planted. . . . [G]razing on State Domain Lands should be discouraged whenever possible."[13] According to state experts, afforestation projects and the habits of goats were in severe conflict. But the goats were not necessarily alone to blame. "The Arab and the goat were responsible for desert wastes," asserted one British governor in the *Palestine Post* in 1934. "It was a misnomer to describe the Arab as the 'Son of the Desert.' He was really the 'Father of the Desert.'" And that desert, the governor argued, was the result of the goat, known from the Old Testament as "an evil beast—a leader in mighty wickedness."[14] Responsibility for the destruction of the land therefore lies with its old inhabitants, both the people and their animals, and the hope for change could be found in newly planted green trees.

Experiences with afforestation elsewhere contributed to the British management of goat grazing in Palestine. "[From] long observation made in Cyprus and Palestine," maintained Gilbert Noel Sale, British conservator of forests, "I have been forced to the conclusion that the old practice of extensive grazing . . . is the prime obstacle to afforestation, soil conservation, and any form of agricultural development."[15] In the eyes of these experts, goats' habits had not only been a local problem but also a threat to the empire as a whole. In order to deal with the problem of goats, foresters in Palestine solicited the advice of an imperial expert, Dr. R. O. Whyte, a member of the Imperial Agricultural Bureau and "eminent co-author of *The Rape of the Earth*"; he noted that "the fact has to be faced that there are to-day no real forests in Palestine and that if there is one country in the world in which afforestation is desirable that country is Palestine."[16]

In other British colonies, foresting policies were not merely rehabilitation or beautification projects, as trees were planted for the sake of firewood and timber, used for construction, heating, and machine operation. In Palestine, however, this was a rather small motivation for planting trees.[17] As a matter of fact, British officials complained, "the production of firewood and timber has long been neglected and ignored by the authorities in Palestine, and, even now, the area devoted to this is totally inadequate."[18] In an area where there was relatively little use for wood for construction or heating, forest planting had other purposes.[19] While the British considered rational land management and expansion an expression of order and control, it was also a reflection of their desire to transform the land and its creatures.[20] In Palestine, a new land order would enable the British to revive a land that they believed was once lush with green forest and plentiful trees—a land of plenty.

According to the British paradigm, goats and other grazing animals damaged nature in Palestine and sabotaged the potential rescuing of the land from its current desolate state (see figure 2.3). Harmful to trees and nature, these animals also hindered other development projects significant to British impressions of progress. In 1942, for example, F. H. Taylor, the district engineer of Lydda, highlighted the damage done to the railway system, one of the greatest modernization projects of the British rule in Palestine. Taylor argued, "The experience on the railway is that since 1936 the damage caused by illegal grazing of goats, sheep, cattle, and camels become [*sic*] more extensive. The first reason is the wave of lawlessness and general contempt for government property, which grew throughout the disturbances, when in many areas ordinary policing died out."[21] The harm to the land caused by grazing animals was thus parallel to and interlaced with the harm caused by local people to the British rule.

As such, the two together—the Arab and the goat—were seen as ruining nature and posing a threat to the governing rule. Taylor's mentioning of 1936, the beginning year of the Arab Revolt in Palestine, is crucial, as this time became a turning point in the regional power structure and political agenda of the British government. Palestinian peasants rebelled against the British rule, growing Jewish settlement, and consequential economic hardships.[22] The British government, although caught by surprise, reacted fiercely to the riots.[23] In the following years, it implemented restrictive partition plans, territorial and others, for the Jewish and Arab populations, affecting their ability to achieve political and economic goals.[24]

Palestinian Arab farmers were hence perceived as threatening to government rule; grazing goats were considered threatening to the land. Harmful in similar ways, these two—the Arab peasantry and Palestinian goats—came to symbolize each other. The need to create order on the land of Palestine was similar to the need to create order among the people of Palestine. Controlling the land and people had become one and the same. To do this, British officials began to record, count, measure, and classify—methods that would allow them to finally eliminate these disturbing creatures and replace them with prolific others.

IF FREED FROM THIS CURSE: NUMBERS ON THE WAY TO REJUVENATION

"Owing to the proximity of the Arabian Desert, too many people are apt to consider Palestine as a natural desert or semi-desert," declared Sale in a lecture he gave to the Palestine Economic Society in 1942. "This view is a

FIGURE 2.3
"Goats Feeding on Wild Vegetation, Kibbutz Sha'ar Ha'amakim, 1935–1940." *Source:* Kibbutz Sha'ar Ha'amakim Archives.

weed, which must be rooted out of all minds. Palestine is a natural garden, and must be restored to its original condition."[25] Sale powerfully believed that if grazing could be controlled properly, Palestine would become a land of plenty again:

> If one valley were to be freed from this curse, its appearance would be totally changed. Floods would be small and rare, if not unknown. The stream would run for several months in a well-defined bed, the banks which would be supported by large undamaged trees. Between the river and the hillside would lie flat, deep and fertile fields. . . . [T]he steeper slopes would be covered with forests. . . . [T]he inhabitants of the village would rapidly gain in prosperity and contentment, and such a rejuvenated valley would be a fair contrast to its neighbors and invaluable object lesson to all.[26]

Not unlike the European travelers of the second half of the nineteenth century, and the Christian missionaries settling in Palestine in the late Ottoman rule (see chapter 1), British officials in Palestine wanted to create a land of plenty.

The first tool used to actualize this vision of rejuvenation was counting. Knowing the number of animals would mean knowing the land, and in this way, help turn the threat of grazing into a treatable problem.[27] Hence the counting of goats began (see figure 2.4). Information about the number of goats was intimately related to understanding the rhythm of Palestinian agricultural life: "Since the milk-yield of the Palestinian goats and the number and growth of kids are so much dependent on the conditions of pasture, the returns of goat-raising vary considerably from year to year. No exact data are available from which to recon the profit in goat-raising since goat-owners do not keep any kind of accounts."[28] The attempt to count goats, by this manner, was also useful in exposing the irrationality at the basis of the local subsistence economy; the lack of standardization and problematic data, which to governmental officials seemed to work against the needs of goat owners themselves, was used as further proof that there was a problem with the way the land had been treated.[29]

According to Sale, the lack of data was joined by structural problems. The existing Ottoman taxation system, he posited, worked in favor of such destructive behavior: "One reason why people without land, capital, enterprise or intelligence adopt and pursue the practice of extensive grazing, is the low tax on goats and sheep. Such a man with a flock of 60 goats can ravage a whole countryside like a conquering army, and can continue to

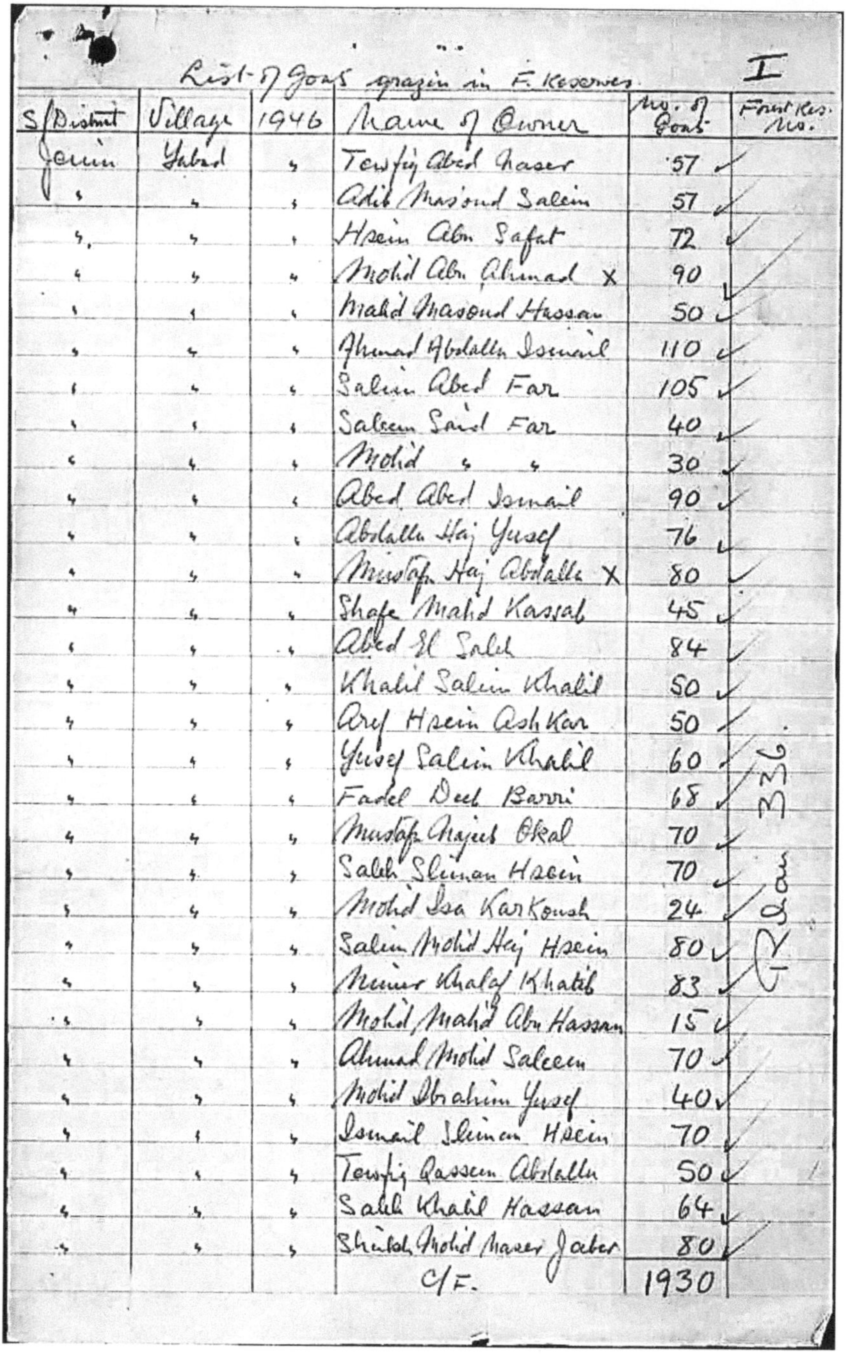

S/District	Village	1946	Name of Owner	No. of Goats	Forest Res. no.
Jenin	Ya'bad	"	Tewfiq Abed Naser	57 ✓	
"	"	"	Adib Mas'oud Salem	57 ✓	
"	"	"	Hsein Abu Safat	72 ✓	
"	"	"	Mohd Abu Ahmad X	90 ✓	
"	"	"	Mahd Mas'oud Hassan	50 ✓	
"	"	"	Ahmad Abdalla Ismail	110 ✓	
"	"	"	Salim Abed Far	105 ✓	
"	"	"	Saleem Said Far	40 ✓	
"	"	"	Mohd " "	30 ✓	
"	"	"	Abed Abed Ismail	90 ✓	
"	"	"	Abdalla Haj Yusef	76 ✓	
"	"	"	Mustafa Haj Abdalla X	80 ✓	
"	"	"	Shafe Mahd Kassab	45 ✓	
"	"	"	Abed El Saleh	84 ✓	
"	"	"	Khalil Salim Khalil	50 ✓	
"	"	"	Aref Hsein Ash Kar	50 ✓	
"	"	"	Yusef Salim Khalil	60 ✓	
"	"	"	Fadel Deeb Barri	68 ✓	
"	"	"	Mustafa Najeb Okal	70 ✓	
"	"	"	Saleh Sleiman Hsein	70 ✓	
"	"	"	Mohd Isa Karkoush	24 ✓	
"	"	"	Salim Mohd Haj Hsein	80 ✓	
"	"	"	Munir Khalaf Khatib	83 ✓	
"	"	"	Mohd Mahd Abu Hassan	15 ✓	
"	"	"	Ahmad Mohd Saleem	70 ✓	
"	"	"	Mohd Ibrahim Yusef	40 ✓	
"	"	"	Ismail Sleiman Hsein	70 ✓	
"	"	"	Tewfiq Qassem Abdalla	50 ✓	
"	"	"	Saleh Khalil Hassan	64 ✓	
"	"	"	Sheikh Mohd Naser Jaber	80 ✓	
			C/F.	1930	

FIGURE 2.4

List of goats grazing by owner, Ya'bad village, Jenin district, 1946, Israel State Archives, Mem-2/4190. Such lists containing information about the number of goats according to owners cover the great majority of Palestinian villages under the British rule.

prevent any form of progress or development over a great area." Ultimately, he argued, "if the tax per head were higher, it would much more in his interest to keep a smaller number of better goats."[30]

Raising taxes was thus a second tool in addressing the problem. It is certainly true that the British government profited tremendously from an organized system of money collecting, as all agricultural taxation skyrocketed in the first two decades of British rule in Palestine.[31] Changing taxation policies in the late years of British rule, however, was more of a strategy to control goats and minimize their perceived damage than a means to make profit. British officials created a differential taxation system such that "the rate for sheep should be slightly higher [than that of cattle], and that for extensively grazed goats at least double or even higher," or in a different report, "cattle and sheep will steadily be favored in preference to goats and camels."[32] The district commissioner of Samaria declared in 1944, "The basic rate on goats to be double that on sheep . . . [with] no progressive rate to be charged on cattle, swine and camels."[33] While regulations varied on sheep and camels, the tax on holding and grazing goats was always the highest. Restrictions grew further, as evidenced by the eventual limitation on the number of goats per owner. The conservator of forests determined in 1946 that "no man should be allowed to own more than 25 goats."[34]

Once numbers were available and taxation policies implemented, recording and standardization were essential. Several documents deal with this aspect of ordering, detailing that "a) a proper enumeration of goats and a proper list of their owners [should be] kept in registers at District Offices; b) only people who have been registered as goat owners at District Offices may keep goats in future; [and] c) only district descendants of such people may inherit the right to graze goats."[35] Goat owners now had to be registered, issued an annual license, and forbidden from selling goats to unlicensed people.[36] The 1946 Shepherds (Licensing) Ordinance determined that "only fit and proper persons over ten years of age will be permitted to graze sheep and goats," and rules would "fix the maximum number of sheep or goats which may be herded at any one time by one shepherd."[37]

To avoid conflict and justify their new regulations, British officials sought to work with the locals, highlighting how information about the state of grazing and affects of restrictions, for example, would be gained through interviews with goat owners. The Committee for the Preservation of Trees and Prohibition of Grazing noted that it would meet with those who had objections to the prohibitions on grazing.[38] The members of the committee added that the money collected from taxes and licensing "shall be used for any purpose designed to assist, aid and educate the shepherds."[39] Beyond justifying the members' actions, such statements reflect a belief that afforestation and limitations on grazing and shepherding would work for the benefit of the land, its people, and perhaps even goats. If raised rationally and according to the guidance of experts, goats would become creatures of high quality and value: "The aim of the Veterinary and Forest Departments," maintained the conservator of forests, "is to improve the breed of goats to a point where they are valuable animals, tethered, fed, and highly productive."[40]

"A MACHINE OF ASTONISHING EFFICIENCY": DEBATING THE NATURE OF THE GOAT

Yet goats were already considered valuable animals to their Palestinian Arab owners, and British actions were understood to be undermining that value. "It is well known that the raising of black goats," detailed an article in *Filastin* in 1947, "is one of the Arab natural sources, and in some cases, the only one." Enumeration, tax raising, and the control of movement posed a threat to the livelihood of both Arab villagers and Bedouins, some of whom relied on the animal's "products and milk as the only source of income."[41] To them, the British regulations and restrictions had nothing to do with assistance and benefit, but instead were a tool to ultimately force Palestinian Arabs to sell their lands to Jewish settling organizations as a last resort.[42] Afforestation projects and the collaboration between British governing powers and Jewish organizations thus focused on the goat as a channel for displacement.

In addition to Palestinian Arabs, some British officials disagreed about the conflicted relation between goats and trees.[43] Several veterinary doctors voiced their grievances against this process and warned about its expected consequences; they positioned themselves as the advocates of goats.[44] The chief veterinary officer, G. S. Emanuel, particularly rejected this vilification of the goat. "A great deal of propaganda has been directed against this animal couched in extravagant terms," he wrote. "It has, among other things, been variously described as a pest, a menace, a black locust with poisonous saliva and a ravaging appetite, etc. etc." Emanuel believed this approach

could harm the people of Palestine: "As a result of this abuse and the sub-
sequent desire for the removal of the goat, the fact is often overlooked
that a not inconsiderable proportion of the population of Palestine depend
upon the goat to supplement their diet. . . . [A] reduction in the number
of animals" will affect "the immediate food supplies of the country."[45] "As
a biologist," wrote G. C. L. Bertram, the chief fisheries officer, to the conservator of forests, "I have long been saddened by the harsh attitude of the
Soil Conservation Board towards the goat, one of the most efficient of all
living machines."[46]

Goats demonstrated tenacity and determination, Bertram argued, elab-
orating on their comparison to machines: "The goat provides the finest
example of assiduity under difficult sturdy toleration of the harshness of
the physical environment, and ability to make something out of almost
nothing."[47] Indeed, raising goats in Palestine and elsewhere was completely
appropriate because "the goat is a hardy paragon of almost all that is desirable in a domestic animal. . . . [It is] the most admirable of all domestic
animals for a poor peasantry. I feel therefore that it in faulty propaganda
on the part of the Soil Conservation Board to try to engender a widespread
belief that the goat is a destructive pest." According to Bertram, responsibility lies with the human race: "The goat like the aeroplane [sic] in war is
a machine of astonishing efficiency which spreads wide destruction when
handled by those who are ignorant or inadequately wise."[48]

Another veterinary officer contended that attempts to restrict grazing
were bound to fail: "A very large force of Grazing Control . . . would be
required for many years to come to prevent the trespass of stock." Additionally, "the preparation of a reliable census of livestock in Palestine as a
prelude to licensing would be fraught with many difficulties." Finally, he
warned, attempts to enforce these restrictions will cause "serious outbreaks
of lawlessness."[49] Even if considered reasonable and powerful, the various
British tools for knowing the land and controlling the goats would not be
sufficient. Such tools of control were also dangerous because these efforts
had the potential of causing greater unrest in Palestine.

State veterinarians therefore chose to subvert the dominant assumptions
about goats and their role in the ruining of Palestine. Reacting to powerful
experts who were, like themselves, appointed officials of the mandatory
state, they attempted to use their own authority as ways of molding plans
of action. By challenging current conceptions regarding the hierarchy of

the creatures of their land, they undermined the rationale at the very basis of the struggle for nature.[50]

British foresters dismissed these objections by linking grazing to political and ethnic tensions: "One of the most frequent causes for breaches of the peace between Arabs and Arabs, and Arabs and Jews at this season of the year is the grazing of animals on lands on which crops have been harvested."[51] Management of forests and control of grazing, through their logic, would also ease tensions between different local communities, which used the land differently, and improve local attitudes to the governing rule. British foresters generally depicted Jewish settlers as a force of progress and rational land management; while data regarding Jewish livestock was lacking as well, they argued that there was no goat problem among the Jews, who preferred sheep and cows.[52]

By the late 1940s, the argument for goats as agents of destruction and call for controlling goat management superseded voices of objection. Despite this, the plans for counting goats along with controlling their movement and reproduction were extremely difficult to execute.[53] Given burdens of taxation and required vaccinations, not everyone agreed to register all the animals they owned (see figure 2.5). Without sufficient numbers and data, British aspirations never fully materialized. Indeed, the British would soon leave the country with the outbreak of war and establishment of the State of Israel, leaving the British goat project unfinished.

THE REPLACEMENT PLAN (OR HOW MANY GOATS ARE WORTH ONE COW?)

Many members of the new Israeli Parliament were mad at goats. In the 139th meeting of Parliament, two years after the establishment of the State of Israel, the first minister of agriculture, D. Yosef, introduced a suggestion for a potential Plant Protection Law, which would work to "prevent the great damage caused in this country in recent years by the herds of goats, which have terminated a lot of the vegetation," and ensure that no man is allowed to "hold goats and herd them except on his own land, and in a number that is secure, [so] that their [female gendered goats] feeding is sufficient."[54] Goats would not only be prevented from entering closed forest areas but also anywhere that was not private property.[55]

FIGURE 2.5

"Arab Shepherds in the [British] Ministry of Agriculture Quarantine Station for Livestock, near Jaffa: Vaccination of Sheep, Goats, Horses and Cows." Photograph by Zvi Oron (Oroshkess), 1934. *Source:* Central Zionist Archives, Image Collection PHO\1356525.

This would enable government authorities to catch goats in forbidden places and sell them.[56] While this was a major aspect of Yosef's suggestion, some members of Parliament objected, not for the sake of goats, but for the private sphere. They rejected the "possibility of using the law to enter private spaces and interfere with house matters."[57] Others argued that Parliament should emphasize that the origins of the law were, in fact, biblical. "I have nothing against this law," Parliament member Eliyahu Moshe Ganhovsky said, "but I object to the introductory explanation. This law is an ancient law in Israel . . . and I think that if we had the fortune to renew it in the state of Israel, then it was worth highlighting that."[58]

In addition to the loss of human lives, dispossession and exile of hundreds of thousands of Palestinians, and loss of their property, the devastation of the 1948 war and the Palestinian Nakba sparked a dramatic reduction in the number of Palestinian goats; estimates of 750,000 goats in 1946 fell to 100,000 four years later, not unlike estimations regarding Palestinian people.[59] In spite of these reduced goat numbers, government officials in the now-called State of Israel quickly expanded formal attempts to control and eliminate them; in 1950, the state passed its first environmental law—the Plant Protection Law, later known as the Black Goat Law.

In the same meeting, Parliament moved to discuss corrections to the British Shepherds Ordinance. "According to the existing law there are arrangements relating to supervising the herding of sheep and goats in a planted area," noted Yosef. "Since the purpose of the law that we just passed to the committee is to forbid the herding of goats altogether in these areas, there is a need to remove the words 'or the goats' from the existing ordinance."[60] Shepherds, members of Parliament agreed, should be licensed and may herd their sheep in designated areas, but they should altogether forget about herding goats.[61]

The Committee of Economics met twice to finalize the new Israeli law, and the chair, Moshe Erem, voiced his concern that the law would be used as "an excuse to bother the Arab."[62] The other members, however, were steadfast: "Known is the great harm caused by the goats, the result of which left the countries of the Middle East with no trees and no shade," said one, and another claimed, "Even the [British] Mandate government in its time attempted to fight this trouble, but with not much success."[63] The Israeli law, with its focus on a system of inspection for illegal grazing, differed little from the British regulations of goat rearing and herding, yet the Israeli law's ultimate objective—the "termination of goat herds while replacing them with house goats, or a herd of sheep, or other means of compensation"—certainly separated it from its earlier British counterpart.[64] One member suggested letting goats graze in desert areas that "do not have trees in them," but the director of the Ministry of Agriculture objected fiercely.[65] "The deserts became deserts by the goats," he said. "The country is desolate because of the Arabs and because of the goats. We got rid of the Arabs and we have the ability to get rid of the goats as well."[66]

The new law was passed, finalized, and handed off to the Ministry of Agriculture as the Plant Protection Law.[67] Two appointed experts, Dr. Pintchi, director of the Sheep and Goats Section, and Dr. Kotzer, director of the Arab Village Section, were in charge of actualizing the law—literally, how to make all unwanted creatures disappear. In their *Plan for Actualizing the Law for the Termination of Goats*, they stated that other, nonharmful animals, such as sheep or cows, would replace the goats.[68]

Similar to the British standardization project, the Israeli replacement plan that followed the Plant Protection Law involved significant counting. Pintchi and Kotzer first estimated the number of goats in the country at 100,000. These were to be killed and replaced with sheep and cows; according to their calculations, 2.5 goats were worth 1 sheep and 12 goats were worth 1 cow. These animals would be available for purchase in neighboring countries such as Turkey, they posited, and the funds for the project (anticipated to take two years to finalize) would come from the money earned by selling the slaughtered goat meat.[69]

As the counting became more systematic, though, it was targeted at eliminating goats owned by Palestinians alone, as indicated by tables with titles such as "Summary of Goat Herds in Arab Villages" (see figure 2.6). Those tables detailing the number of animals owned by Arab owners or entire Arab villages reflect the plan's goal. In Beit-Jan, the village containing the greatest number of goats according to the 1949–1950 census, for example, 3,394 goats were indicated as registered and exactly 3,394 goats were indicated for termination; the goats would be replaced and "rehabilitated" by 850 sheep and 170 cows—a logic that was applied to all other Arab villages in the country.

While these documents clearly outline the law's plan, it is less clear how well these plans materialized. Despite a goal of two years, correspondence in later years indicates delays along with an overwhelming amount of remaining work and planning. Officials appeared to be willing to use the power granted by the law to confiscate illegally grazing goats, but they were less willing to kill them. Miscommunication of marketing and pricing contributed to a delay in "goat replacement." When a private agricultural produce trade company notified participants that the Central Cooperation for Agricultural Production in Palestine (Tnuva; see chapter 4) had agreed to "pay for slaughtered meat—[at] 40–45% of the weight of the living animal—1.5 Israeli Lira for each kilogram, meaning about 700 pruta for 1 living kilogram," it also cited confusion over how to "know by which of the two prices I should base my calculations when I come to get the goats from their owners in exchange for the sheep they receive."[70]

Moreover, although the execution of the replacement plan certainly depended on the availability of sheep and cows, it relied more on villagers' consent. The Israeli Ministry of Agriculture initially considered the use of military power, but this attitude was, at least officially, changed.[71]

סיכום ערו"ד העדרים בכפרים הערבים לעונה 1949/50

.פס נכה לשקום	.פס כבשים לשקום	.פס עיזם לחל	מה"ב פקנה	.פס פרות שוריא	.פס כבשים	.פס עזים	סח'יצ אדמות מרעה	שסח מרעה בר. גב.	שסמ גורלי שדה בר גב.	ה כ פ ר ו
										גליל מערבי
15	175	504	789	191	94	504	9597	2697	6900	1. אבו סנאן
50	440	1370	1874	380	118	1378	26686	12686	14000	2. עראבה
50	485	1471	2052	350	231	1471	11157	4857	6300	3. גולים
40	365	1130	1243	113		1130	5038	4118	920	4. ג'ת
25	250	754	1006	252		754	12369	7269	5100	5. דיר-חנה
13	100	334	522	173	15	334	7916	5666	2250	6. דיר-אסעד
75	500	1744	2077	329	4	1744	25751	18901	6850	7. ירקה
95	500	1945	2192	247		1945	11810	10000	1810	8. יבנה
25	200	651	711	60		651	3565	135	3430	9. כאוכב
15	175	501	923	251	71	501	7276	3676	3600	10. כבול
47	300	1074	1240	166		1074	6697	4847	1850	11. כפר-סמיע
24	175	589	831	242		589	13715	9465	4250	12. מגדל-כרום
76	500	1768	2145	325	52	1768	18371	14621	3750	13. מכללה
20	200	697	1162	465		697	30119	19319	10800	14. סוג'הר
18	150	489	856	232	139	485	12901	9951	2950	15. נחף
95	500	1944	2837	858	55	1944	53386	37024	16362	16. סכנין
13	150	430	648	106	112	430	1874		1874	17. סגור
60	75	756	1139	383		756	22780	18280	4500	18. פסוטה
125	500	2255	2927	560	132	2235	29145	14345	14800	19. תמרה
10	150	594	749	287	68	594	13640	5140	8500	20. זילבון
14		141	226	28	57	141	370		370	21. תרשיחא
905	5830	20925	28059	5936	1148	20925	324163	202997	121166	ס ה " כ
										גליל עליון נ ה ח י
170	850	3394	3512	113		3394	21732	15591	6141	22. ביר ג'ן
30	195	689	1065	294	82	689	6650	4340	2310	23. ג'יש
50	250	1033	1426	233	160	1033	15344	11294	4050	24. חורפיש
45	250	941	1157	143	75	941	8403	5403	3000	25. פקיעין
20	100	414	580	165		414	13547	12257	1290	26. רה סם
10	35	168	747	205	168	374	2616	560	2056	27. ריחניה
325	1670	6845	8487	1153	484	6845	68292	49445	18847	ס ה " כ

FIGURE 2.6
"Summary of Goat Herds in Arab Villages in the 1949/50 Season," in T. Kotzer, director of the Arab Village Section, to A. Ben-David, general secretary of the Ministry of Agriculture, "Termination of Herds of Goats in Arab Villages," November 23, 1950, Israel State Archives, Gimel Lamed-19/17022, 1. The heading of the ninth column from the right says, "Number of goats for termination," which is an amount equivalent to the overall "number of goats" in each village. Thus according to the plan, all goats would be terminated.

"We intended to begin executing the termination of the goats," wrote M. Hanuki of the Division of Agricultural Department at the Ministry of Agriculture, "but we postponed its beginning until we can offer them appropriate compensation. We are taking care of getting sheep and cattle so that we can offer them in exchange for the goats. Until then, we do not intend to begin with the plan, except in those cases where the Arabs willingly agree out of their own good will."[72] According to S. Zamir, the officer of development of the Triangle Area in 1954, for example, Palestinian villagers were themselves "asking to replace the goats with sheep."[73] The Division of Agricultural Development at the Ministry of Agriculture then followed up, asking him "to find out the conditions and ways in which the replacement can take place."[74] Thus while officials were certainly determined to eliminate the goats, they nevertheless realized that the project depended on negotiation.

As the years passed, the burden of negotiation became more apparent. The Arab and the goat continued to pose a challenge by crossing borders, entering forbidden areas, and petitioning the state. They too, like the contrary voices of the veterinary doctors during the British rule of the 1940s, were challenging official plans and official people. By as late as 1958, the Black Goat Law had not been enforced or executed. Fantasies about order remained unfulfilled, highlighting the fragility and anxiety involved in controlling the land along with its animals and people.

"INSEPARABLE PART OF THE SUSTENANCE OF THE ARAB FELLAH": PETITIONING THE STATE

The reactions to the British restrictions varied, and so did the responses to the Israeli Plant Protection Law. While some goat owners simply ignored official limitations in a manner that anthropologist James C. Scott termed "everyday forms of resistance," others attempted to change official decisions by way of petitions.[75] Petitioning the authorities was a rather common practice and in fact had been a major way of communicating with the governing body for a long time.[76] Much like the Ottoman rule that preceded it, the British Mandate in Palestine was designed to have minimal interaction with the daily lives of farmers and pastoralists as long as they complied with the rules. While the pressures of the state on Palestinians grew with the creation of the Israeli military rule in 1948, the villagers and nomads remained distanced from the attention of the state on the grounds that they stayed within the designated geographic limitations.[77] Given these distant modes of governance, farmers and pastoralists become most vocal at times of violation of such state of affairs, such as when rules changed or taxes peaked.[78] It is during these moments of state interference that many attempted, rather creatively, to react to the governing rule with the tools of the state already in place.[79]

The Israel State and Israel Defense Forces Archives both hold dozens of passionately written petitions that deal with goats and grazing restrictions. Most were sent by pastoralists and villagers, with some handwritten, a couple printed, and a few signed with a thumbprint (see figure 2.1). While many petitions were addressed to British officials and plead for changes in grazing restrictions, the majority called on Israeli officials as the rules became stricter and more focused on eliminating the goat population entirely. In spite of variations in style and form, all the petitions highlight the contrast between the local understanding of the value of the goat and governmental demands to restrict its habits and existence.

Petitioners were aware that the vilification of the goat and limitations to its grazing had political motivations. The leaders of the Esh-Shibli Bedouin community, for example, wrote in 1946 to the British commissioner in Jerusalem about the gradual loss of their land to Jewish projects:

> Some time ago an area of about 3000 dunams was taken from our lands and given to [the Jewish] Khadoorie Agricultural School, and another area of 3000 dunams was given to the Palestine Jewish Colonization Association, and 3000 dunams remained as grazing lands for our Cattle and Flocks on the East-northern side of Mt. Tabor and we were promised by the Government that this area will be always used as Grazing lands for our flocks.[80]

Despite these protests, the British government would soon seize more of the Esh-Shibli land: "Three months ago Forest Department Officers came to the tribe and wanted to dig holes for erecting iron marks and barbed wire for the purpose of making those lands Government Reserved Area, but we have prevented them as these land are the only grazing area for our flocks and had at that time submitted complaints to this effect, and up till now we received no reply."[81] A sense of acute violation of the status quo brought the people of Esh-Shibli to use all ways possible, including petition writing and physical resistance, to prevent the repossession of their lands and disregard for their ways of living.

Similar petitions were sent to the Israeli government after the establishment of Israeli rule in 1948. The goat owners of 'Ein-Sahala village, for instance, argued that the prohibitions on grazing and holding goats were like "taking the last piece of food out of our mouth." Because 'Ein-Sahala existed on mountainous lands that did not allow for sufficient crop growing, the villagers' only source of living was the goats. Villagers highlighted the significance of goat milk to everyday life, noting that "the entirety of our food in these hard times of bad nutrition is the milk of goats." They objected to the government's decisions, and demanded "reconsideration of this matter since we are Israeli residents with democratic rights, and we do not want our rights to be damaged, or to be subjected to abuse and racial persecution."[82]

Beyond grazing restrictions, the animal tax was a main concern of the peasants who sent petitions to government officials. "We," wrote the goat owners of Sakhnin village to the minister of agriculture in 1954, "herd our goats on our own land . . . and pay enormous amount of taxes for them. We did not disobey the order of the Forestry Department at all . . . [but] this year we were ordered to pay 200 pruta for each goat, and this payment is too heavy a burden."[83] Many petitions addressed the issue of increasing the annual tax on goats, especially those coming from the Druze village of Beit-Jan, which had the biggest number of goats under Israeli rule (see figure 2.6).

Druze villagers were some of the most prolific petitioners, particularly after the establishment of the Israeli military rule, when the number of limitations on goat practices especially increased. Of the non-Jewish populations living in Palestine at this time, the Druze people were understood as supporters of the Zionist cause (as some supported Zionist forces during the 1948 war, and others served in the Israeli Defense Forces prior to becoming citizens of the State of Israel)—a support that should be understood in the context of state attempts to use ethnic and religious minorities strategically to dismantle the power of local majorities. The Druze political stance aligned with a relative trust in state institutions; a growing contrast between their expectations and the government's damaging policy, combined with the belief in the functionality of state institutions, made their petition writing worthwhile. Such were the efforts of one Druze village called Beit-Jan, which became a focal point for petitioning throughout the 1950s. The people of Beit-Jan as well as those from the surrounding region complained about the rise in goat taxation and repossession of their grazing lands as national forest reserves.[84] Like the Esh-Shibli case, the Beit-Jan campaign was especially strong given that the community had come to rely on goats as its main source of living. Ultimately, the Beit-Jan protests made their way into the court and daily papers. In winter 1954, one article announced "a severe dispute between the Druze and the Ministry of Agriculture regarding the grazing of goats," and observed how even though many other villages, Christian and Muslim, had begun terminating their flocks, "the Druze villagers, that felt very strong, rebelled and refused to pay the tax and even sent their flocks to herd in government forest without permission."[85]

Most often (and not unlike veterinarians of the British rule), petitioners concentrated on challenging the logic behind grazing limitations, suggested replacement plan, and idea that the goat was harmful to the land. Some—as did the people of A'ara village whose petition opened this chapter—used theological reasoning to argue that goat rearing was beneficial and natural to the land, seeking to juxtapose governmental actions with godly intentions. Other petitioners presented the utility in raising goats. Such was a petition from villagers of the Nazareth area to the chair of Parliament in 1952; reacting to the Plant Protection Law, these petitioners decided to "spur mercy." "We used to send milk in great quantities to dairies in Israel, that in addition to the benefit we got from the animal droppings as manure for our fields. In raising goats there is much blessing to our region and there is no expected damage what so ever."[86]

The majority of petitions discussed not only the use of the goat but also its necessity. "The fellah," wrote the people of Sakhnin village, "from the nature of his life needs to hold goats and sheep in order to produce benefit from their products, wool, and waste." The goat was thus a natural aspect in the life of the fellah. But it was part of the nature of the land too:

> The majority of the land of our village Sakhnin is a rocky land, high mountains and valleys . . . and the goats are the only kind of livestock that can exist and reproduce on such lands. The nature of the land itself, therefore, requires that we hold goats rather than any other kind of livestock. . . . [G]oat raising is an inseparable part of the sustenance of the Arab fellah, and in its dismissal is a destruction to one of the assets of his life.

Just like experts and state officials, farmers made claims about what was natural and essential to the land. Because the State of Israel has "a guarantee for the livelihood of the Arabs within the borders of the state," it also had to, they argued, acknowledge this triangular bond between the goat, the fellah, and the land.[87]

Furthermore, some petitions challenged the logic at the basis of the replacement plan itself. The owners of goats living in the Baqa al-Gharbiyye area, for example, asserted that while they are loyal citizens, the plan does not make sense according to the temporality of goat rearing. The replacement plan was to take place in the early spring, a time when the kids are born and milking season begins. "How could the owners sell or slaughter [the goats] when they are thin and unworthy of being eaten?"[88] They instead suggested waiting until the summer months and when the government is able to substitute goats with other livestock.

While it is unclear whether any of these petitions influenced official decision-making regarding goats, they certainly did not go unnoticed. Both British and Israeli officials translated and circulated these petitions among themselves. One handwritten note attached to a petition from the village of 'Ilabun, for instance, explained in Hebrew that "the inhabitants of the 'Ilabun village complain about the governmental intention to slaughter and terminate the flocks of black goats and give explanations for why these goats are not harmful, and on the contrary—that they are even beneficial."[89] "Please see the request of the goat owners of the Baqa al-Gharbiyye area," noted an official of the military rule, and "examine the possibility of agreeing to it."[90]

Considered together, these petitions addressed different aspects of land-ownership, animal grazing, taxation, and agricultural life and temporalities. In order to win the support of their readers, petitioners tended to compliment state officials and the state while simultaneously highlighting the injustices of their legislation. In most cases, they paint a picture of deprivation and discrimination, but most clearly, they speak to a message of incongruity. To goat owners, the government was endorsing a senseless way of treating the land. By writing to the state, they demonstrate a belief in their own power to change these ways, and if one judges from the failure to enforce the law and replacement plan, they were not entirely wrong.

NO PROGRESS WITH MIXTURE: THE RACIALIZATION OF GOATS

Not only Arab farmers and goats found limitations on grazing troubling. While the official British and Israeli agenda considered goats a threat to the

FIGURE 2.7

The goats that Europeans found in Palestine: bad goats. *Left*, a local, black herding goat. *Source:* Ag. Moshe Schorr, *The House Goat* (Tel Aviv: Hakarmel [published by the chief supervisor of agricultural education, Ministry of Education and Culture], 1949), 14. *Right*, a Damascene, brown goat. *Source:* Dov Beker, *Sheep and Goat Rearing* (Ein-Harod: Hebrew Shepherds Association, 1948), 67.

revival of Palestine, many Jewish farmers celebrated goats for their important role in the growing Jewish settlement. Moreover, not all goats were the same. Although Jewish farmers held local goats from the earliest days of Jewish shepherding (see chapter 3), they soon learned that there were some good goats and some bad goats (see figure 2.7).

Jewish experts were invested in measuring and classifying the various kinds of goats that they found in Palestine, and ultimately encouraged raising the house goat, which was an imported breed (mostly from Switzerland and Romania).[91] "The raising of house goats has spread in all civilized countries and is supported there by governments, municipalities, and social institutions because they see in the raising of goats something that would do much for easing their situation and improving the nutrition (feeding) of the masses," contended livestock experts of the Jewish community during the British rule, pointing to the the United States and Russia as examples. "This is while in our country still prevails an argument about the value of goat milk, and there are still fastidious people who think it has 'a special and unpleasant smell and taste' (by the way, if the goat and the pen are kept clean, these accusations have no basis)." House goats were the best fit for the intensive nature of the growing Jewish settlements, particularly those semicommunal settlements, the moshavim. "In our country," the experts continued, "where most of the farms are built on very limited stretches of land, and now that we are on the verge of a huge settlement movement, there is a great future in raising goats as the basis for the small farm."[92] In 1953, leading Israeli farmers established the Association for House Goat Growers (Aziza) in order to promote the proper rearing of house goats for locals and new immigrant farmers (hailing mostly from the Middle Eastern and North African Jewry).[93]

The basis for handling goats in small Jewish farms was the assumption that goats—even European house goats—have an innate capability for being destructive, not only to forests, but the pen: "As is well known, more than other beasts the goats is more capable of ruining and wasting the food of which it is served, if the stall is not organized properly."[94] Farmers were taught how to raise house goats rationally through explanations, images, and drawings (see figure 2.8), with the hope of managing the "capricious nature of the goat."[95]

As opposed to local herd goats, house goats were raised in small numbers, fed in a pen, and produced more milk. Most important, house goats were not meant to graze, as experts believed that the vegetation could harm their delicate bodies and udders.[96] "While the Swiss Saanen goats"—which became the dominant breed on Jewish farms—"are capable [of] climbing the Alps," said one guidebook, "they are fine being tied throughout their lives."[97] In addition, although British authorities often referred to the local herding goat as "the poor peasant's friend," Jewish experts determined that house goats should be regarded as "the poor farmer's cow," or just minicows.[98]

House goats were not only prolific and European but white too. Because the Plant Protection (Damage of Goats) Law explicitly stated that the "prohibitions do not apply to house goats, of which there is no expected danger to plants," the law came to be known as the Black Goat Law.[99] Therefore with the passing of the law, Jewish experts decided to concentrate on the problem of herds of sheep and goats that continued to coexist and blend on Jewish farms. In November 1950, the Hebrew Shepherds Association stated in a meeting that members were not "goat haters" but instead believed that the mixing of sheep and goats should be banned. "There is not room for linsey-woolsey; there is no room for holding goats and sheep together," they said, arguing, "Herds cannot make progress with mixture." And as for "the issue of the Goat Law: The report explains that the law does not refer to our 1,000 [white] goats, and that there is legal agreement to hold them."[100] Mixing threatened governance.[101]

משרד החינוך והתרבות
הוצאת המפקח על החינוך החקלאי

אגר' משה שור

עז הבית

FIGURE 2.8
A good goat. Cover of Ag. Moshe Schorr, *The House Goat* (Tel Aviv: Hakarmel [pub-lished by the chief supervisor of agricultural education, Ministry of Education and Culture], 1949).

Experts decided not only to separate sheep from goats but also white goats from black goats and Jewish goat owners from Arab goat owners. "With the exodus of the Arabs and the Bedouins from the country, in the days of the War of Liberation [referring to the 1948 war] it seemed that the days of the goats in Israel were over," noted one daily newspaper, *Ma'ariv*, in 1953. "The local black goat, the one growing horns, that same 'devil,' which is known as the cruel enemy of every tree and blooming plant—there is almost no trace of her [*sic*]. Only . . . some small flocks have survived.

Until the exhibition of Aziza was arranged, only few knew that the goat has reached greatness again among us, and that in our country there is a flock of more than 20,000 heads." "But please," it added with a hint of humor, "do not compare these aristocratic, polite goats, of a pure Swiss breed, with those goats . . . even the billy goats—they are so respected, dolled-up, and their beard is long and white. Some of them have a handsome forelock that flutters above their intellectual forehead."[102] According to agricultural experts, white house goats could be admired and celebrated, but black goats were to be castigated and demonized. Using ecological, racial, and national reasoning, Israeli officials discouraged Jewish farmers from holding black goats: "From a national stand point," concluded those at the Ministry of Agriculture, "we are not allowed to fight the war of the 75,000 Arab goats that terminate every bush and grove."[103]

The racialization and ethnicization of goat management continued further. State instructors soon began to pursue the separation of herds in Jewish farms and, more vehemently, attempted to convince Jewish farmers to terminate their nonwhite goats. Haim Schwartz, one such instructor to a series of southern settlements, summarized his experiences in a report from 1955, observing,

> In Mash'en [village]: The big number of goats that exist in the village does not bring any blessing but I don't see any way to convince them to get rid of this harmful branch. . . . [I]n the meantime I managed to separate the goats from the sheep. In Ruha [village]: a big problem. . . . [M]any people in the village started buying Arab goats in great numbers from dubious sources. I was not able to convince the members not to take the sheep with the goats to graze.[104]

As the government's middleperson, Schwartz's attempts at implementation highlight the disparity between governmental policy and practices on the ground. To the government, goat admixture had stopped making sense, and had become threatening to the new racialized logic of nature and order, but to those living with goats, the new separatist way of thinking misaligned with long-considered conventional knowledge and conduct. Ultimately, many of these goat owners sought to understand goats as they had before.

NATURAL SYMBOLS

There is a long history to the relation between the image of the goat and the understanding of Palestine/Israel. According to the work of Palestinian

Arab doctor and ethnographer Tawfiq Canaan in the 1920s, goats and other animals were central to Palestinian demonology. Goats were among the preferred forms of appearance for demons (*djinn*) in Palestine, partially due to their black color. While not considered the most insidious among animals (a role assigned to the hyena), "a goat stands," Canaan posited, "for a bad demon."[105] There is also longevity to the relation between goats and the notion of the Holy Land—one colored by the biblical idea of the scapegoat and its varied, often negative meanings. This is exemplified by *The Scapegoat*, an 1856 painting by William Holman Hunt in which he attempted to depict the biblical environment. Armed with research from two weeks in the Dead Sea as well as two goats, a full-length skeleton of a camel, and a skull of an ibex, Hunt drew Jesus as a goat, dying in a horrible, lurid wilderness.[106] Although the painting attracted much attention, leading art critics in London described it as a complete artistic failure, with some even denigrating it as a disgraceful representation of the Holy Land.

Many decades after Hunt, British and Israeli experts similarly interpreted the Palestine/Israel environment and analyzed its creatures according to biblical notions. Black Palestinian herding goats fit neatly into the biblical idea of hairy devils: "The main agent in the execution of the curses" that this land was subject to since antiquity, explained Sale, the conservator of forests of the British government, "has been the goat, fitting symbol of all that is devilish and futile. . . . The goat population, taken as a whole, succeeds in revenging the miseries of its members by impoverishing the human race."[107]

Along with biblical notions, government and settler understandings of goats and grazing in Palestine along with the legal practices that accompanied them—be they successful or not—were entangled with particular ideas about race, control, and rationality. Black goats came to be seen as destructive, rebellious, and Arab; white goats were productive, polite, and Jewish. Such practices were also woven into the debate about what was "natural" to a land that many people wanted to transform. To the governing rule, goats and Palestinian Arab peasants gradually became one and the same; the legal system was used to fight against both for the sake of environmental and political goals, even as its tools remained insufficient.

The coupling of goats and Palestinian peasants is remarkable considering the long tradition of coupling goats with diasporic Jews. In Britain, for example, the goat became a frequent symbol of the British other, the

Jew, in the late nineteenth century. A cartoon in Britain's *Fun* magazine,
for instance, depicted Hunt's contemporary—Benjamin Disraeli, who was
mocked for his Jewish origins—as a (devilish) goat.[108] More salient, however, was the coupling of the Jew with the goat within Jewish folklore, particularly Yiddish folklore. In numerous examples, the goat symbolizes the
Jew and Jewish village life; the scapegoat represents the persecuted Jew;
and the milk of goats connotes the relation of the Jew to the Holy Land.
Both Sholem Aleichem's *The Bewitched Tailor* and Isaac Bashevis Singer's
Zlateh the Goat, for example, tell stories of Jews and goats with marvelous
powers.[109] Similarly, *The Tale of the Goat*, a fable that was widespread in Jewish communities around the world and finally published by Shmuel Yosef
Agnon in 1925, tells the story of a miraculous diasporic goat, who traveled
to the Holy Land through a secret tunnel in order to return to their Jewish
owner, and carry milk that was plentiful and "sweet as honey" (see figure
2.9). Contemporaneous paintings include many of Marc Chagall's pieces as
well as Reuven Rubin's early works in Palestine.[110] Rubin's coupling of the

FIGURE 2.9
Cover of Shmuel Yosef Agnon's version of *The Tale of the Goat* (Jerusalem: Hagina
Publishers, 1925), painted by Ze'ev Raban of the Bezalel School of Art in Jerusalem.

Jew (and himself) with the goat is perhaps one of the latest examples of this tradition. With the growing European settlement project in Palestine, goats ceased to be a symbol of the Jews and their relation to the Holy Land.

As part of the process of the European settlement, natives to the land of Palestine/Israel were marginalized and blamed for the destruction of the land. Different historical actors, however, be they state officials or farmers, Arabs or Jews, humans or animals, attempted to resist this process of scapegoating in various ways, and challenged the desire to transform the land and its creatures. But the power of that desire prevailed. The establishment of the Green Patrol in 1976 under the Israel Nature and Park authority, for instance, intensified state control over people and animals whose movement and ways of living were deemed illegitimate.[111] Not only were black goats and Bedouins marked as enemies of nature but recently camels were, too.[112]

It is important to take note of one final, more recent change. In the last two decades, the scientific paradigm regarding goats and grazing has shifted. The global scientific community now perceives goats as necessary to the balance of nature—specifically for preventing forest fires.[113] Today, experts in Israel link the steep rise in forest fires to the elimination of goats, and in a surprising turn of events, the Ministry of Agriculture has begun to use incentives to convince Bedouin shepherds to bring their flocks to herd in the Carmel and Jerusalem forests. Perhaps not surprisingly given this story of denunciation, this has not met with much success.[114] Lastly, following a vote in the Israeli Parliament, this shift in the understanding of goats became manifested in the law: as of May 13, 2018, the Black Goat Law had been canceled.[115]

3 THE RISE AND FALL OF HEBREW SHEPHERDING

"Your letter hit me like lightning from the sky," wrote pharmacologist and hormone researcher Felix Gad Sulman (1907–1986) to his friend and colleague Aaron Harari (1908–1984) at Kibbutz Merhavia in 1962.[1] "I can understand all the reasons but the act reminds me of a man that would divorce his wife because . . . she served him cold soup." Clearly stunned, Sulman concluded his letter with a substantiated backing: "I support your struggle. I am certain that with joint forces a solution to this painful problem can be found."[2]

The problem in question? The potential closure of Merhavia's sheep pen. Despite Sulman's hopes, no solution materialized and the pen was closed for good, leaving shepherds in search of a new occupation, Sulman's experiments unfinished, and the sheep homeless. Surprisingly, however, Sulman's reply was not the most dramatic reaction to the kibbutz's decision: David Zamir (1906–1967), the leader of the shepherding community in Israel, was devastated, and a few years later, committed suicide at the kibbutz.[3]

The closing of the pen at Merhavia was not a unique event. Ultimately, pen closures became part of a contemporary trend in Israel, leaving few independent sheep farms to handle the small sheep cheese market and limited (but growing) demand for sheep meat. The end of a story that began at the turn of the twentieth century, sheep herding among Jewish settlers emerged in the last years of the Ottoman period in Palestine and then prospered during British rule, until pens existed in the majority of the different Jewish agricultural settlements (the kibbutz and moshav communities) with the establishment of the State of Israel. This chapter describes the rise and fall of Jewish sheep herding in Palestine/Israel, and hones in on the relation between settling practices and the settlers' ever-changing means of understanding the land.

The European manners of settling in Palestine were entangled with shifting possibilities of landownership and use. Gradual Ottoman changes in land tenure and the emergence of private property, which both received their legal stamp in the Land Code of 1858, culminated during the British and Israeli rules. As a result, the land and its creatures ceased to fill the multiplicity of roles they previously held. Each plot of land had a permanent use now, and each animal had one purpose. Ownership of land, now coupled with documentation, gradually moved into European hands, creating a starker contrast between nomadic and sedentary life.[4] The commonly practiced local system of collective ownership—the *mushā‘*—was deemed illegitimate, although this happened along with the growth and global admiration of the collective use of land in Jewish agricultural settlements.[5] The way to know and treat the land changed as well; in addition to exploring the width and length of the land, it became a time of searching deeper to unearth biblical truths from the underground and pump water. Intensive agriculture, comprised of different kinds of knowledge and skills, increasingly gained prominence.

In the midst of these changes, Jewish settlers debated the right ways to become native to the land and correct approaches to becoming true Hebrews. Knowledge of the land of Palestine, which they equated with the land of the Bible, was considered essential to becoming part of land, for belonging to and owning it. The means to explore the land and its creatures were numerous. Some focused on mapping, specimen and data collections, and digging; others centered on bodily ways of knowing.[6] Although just one of many methods, the Zionist practice of walking the land, known as the *tiyul* (the "stroll"), became widely used and institutionalized.[7] Another was shepherding and the rearing of sheep—a practice that settlers considered to be the most ancient and biblical of all, or the so-called crown of tradition.[8]

This chapter deals with two main Jewish shepherding movements, Haro'e and Hanoked (both meaning "the shepherd"), from the first years of the twentieth century until the early 1960s. These groups shared a common goal of redeeming the land of Palestine though shepherding, although they disagreed about the path of this redemption. Some shepherds used their senses, and others used numbers; some walked and shouted in the open spaces, while others danced and sang near the pen. I examine these different approaches to shepherding in the land of Palestine/Israel and explain why shepherding is a largely forgotten memory of the past.

THE NEW-ANCIENT HEBREW SHEPHERD

"It was in 1920 at the end of the summer, in my first months in the country," recounted a settler named Efraim Eliash. "I was working near . . . the mountain, and suddenly heard the sound of bells. . . . I turned my face and . . . a flock of sheep was sliding towards me, and behind it the shepherd—a very peculiar figure. . . . [I]t was a remarkable image. The figure of the shepherd, the flock and the mountain slope, all this impressed me and I was enchanted. This was the first time I met a Hebrew shepherd."[9] Eliash's encounter ultimately shaped his decision to adopt shepherding himself. His deep fascination stemmed from the fact that such a sight was extremely rare. While sheep and shepherds were common in early twentieth-century Palestine, and shepherding a main occupation for some Arab peasants and most Bedouin pastoralists in the region, the vision of a Jewish shepherd in Palestine was for a long time nothing more than a European fantasy.[10]

European settlers in Palestine, and Jews especially, perceived an inherent fit between sheep and the land. Sheep, much like cattle, were a crucial part of the agricultural economy in their places of origin.[11] More important, however, sheep were seen as the most natural inhabitants of the Palestinian environment since Europeans equated Palestine with the land of the Bible, and many of the leading figures in the Old and New Testaments were, in fact, shepherds. "It is enough to browse through the Bible to realize the extent to which our fathers were dealing with sheep. The shepherd and his herd are mentioned over 300 times," wrote Zamir, many years before his tragic death. "It has been proven scientifically that . . . at the birth of our nation, raising sheep was the main occupation of the Hebrews," he added, noting the great impact that shepherding had made on generations to come. Biblical shepherding deeply influenced Christian thought and practice as well, he argued, observing that "many biblical symbols originate in sheep herding. The shepherd, who became an exemplary figure, is the source for the Christian term pastor." Moreover, Zamir posited that raising sheep was not only natural to people of the Bible but also natural to this land. The choice of sheep was "not a result of some vague affection for sheep rooted in the Hebrews. The natural conditions of the Land of Israel are convenient for the raising of sheep. The entire region has always been blessed with herds."[12]

In order to become true Hebrews in the land of the Bible, then, some Jewish settlers wanted to be shepherds, yet they knew little about caring for sheep and goats. Ultimately, Jewish settlers' solutions to the knowledge gap played into a contemporary understanding of local peasants and pastoralists. In the early twentieth century, European travelers, researchers, and settlers as well as some Palestinian intellectuals considered the local Arab population to be the link between the biblical past and modern Palestine, and therefore they studied and to some extent adopted Palestinian Arab ways of life.[13] In following local practices, Jewish settlers hoped to unearth the link between the people of Palestine and the Hebrews of the Bible. By "going native," they tried to create a new-ancient man, a modern Hebrew person.[14]

Unlike agricultural settlers of the late nineteenth century, or those living in the cities, Jewish settlers seeking "native Hebrew life" attempted for a while to be pastoralists like the Bedouins.[15] Many lived for long months with Bedouin tribes, adopting particular practices such as the nomadic style of dressing and knowledge of the local Arabic dialect. They sought not only knowledge of the habits of the land but also the creation of a Jewish Bedouin nomadic tribe.[16] Since Bedouins were considered the masters of shepherding, settlers recognized the need to learn the practice firsthand from them. Ultimately, in living with Arab shepherds, Jews learned to handle and love the sheep (and goats); in particular, they came to understand how to feed, milk, shear, and care for the sheep, and most important, control the sheep out in the open.[17]

Controlling the movement of sheep was not an easy business. Only skilled shepherds could successfully direct the sheep to a desirable grazing area, control their good pace and orderly manner, and convince the sheep to return in case of danger or bad weather. Knowledge of shepherding included the ability to differentiate between the individual animals of the flock and communicate with these different members. One shepherd described how, while training with Bedouins, he came to appreciate the individual sheep, not just the crowd; indeed, shepherds learned to recognize individual sizes, colors, types of hair, lengths and shapes of the ears, and differential "facial expression[s], behavior[s], and characteristics." Beyond these features, there were "signs that [were] hidden, that only the sharp eye of an experienced shepherd [would] notice. . . . [Y]ou realize that the sheep is not so innocent as you initially imagined, that she has her own wisdom, and that she is not so helpless and miserable."[18] Recognition and differentiation went beyond the training of the eye. The shepherd and sheep learned to recognize each other by smell as well. "The Arab has a special smell and the sheep smell it," wrote a different Jewish shepherd. The connection went both ways, as shepherds could also reportedly smell differences "when two herds were mixed."[19]

In addition to sights and smells, sounds were an important way of managing the movement of sheep in space. Many Jewish settlers noted how Bedouin shepherds used their voice to manipulate the sheep, describing a near-magical use of vocal control. One settler recalled witnessing a Bedouin shepherd "calling his herd, after they went too far. . . . [H]e was shouting and doing magic tricks with his entire body." When the Bedouin shepherd noticed his spectator, "his enthusiasm grew, and he threw himself to all directions, his head-cover flew off . . . [and] he was waving with his *abaya* [dress] . . . falling and standing up, screaming and stopping and screaming again. He was all 'hocus-pocus.'"[20] Though a generally mystifying view of Indigenous people was typical, Jewish settlers did care about the particularities of the shepherd's bodily sounds and movements because they considered the Bedouin shepherds an authority. Only by knowing how to use their voices and move their bodies like Bedouins could Jewish shepherds master their flocks, become Hebrews, and redeem the land. For instance, it was told that Kozchuk, one successful Jewish shepherd, managed to draw "the flock after him as hypnotized, with wonderful Bedouin calls."[21]

The Bedouin habitus was not always enough, however, as the Bible and its depictions also played a crucial role. Moshe, a shepherd from the Kinneret settlement, highlighted the use of biblical words: "We studied the Bible and were enthusiastic about descriptions of the nomadic life. We used several biblical expressions, which were missing in our professional work. We liked the biblical stories about the wandering tribes in particular. Because we were shepherds!"[22]

The sheep reacted well to this amalgamation of Bedouin practices and biblical language. A witness, enchanted by the musical impact of one Jewish shepherd, once observed, "They [the sheep] turn their heads towards the shepherd and listen, listen and enjoy." The witness noted the shepherd's "curly wild hair that fell on his eyes," and the way in which he stood "in the middle of the herd, raise[d] his head up and beg[an] to sing a Hebrew song with an Arab tune." The witness continued, describing how "he finished singing one song, and began a second song, using a wild Bedouin music: A l-a-n-d f-l-o-w-i-n-g w-i-t-h m-i-l-k a-n-d h-o-n-e-y."[23] Thus emerged this first generation of Hebrew shepherds; they named themselves Haro'e (meaning "the shepherd," in biblical as well as modern Hebrew).[24]

While training with Bedouins was considered essential to shepherding, "going to the Bedouins" was by no means a widespread phenomenon. In the end, the number of Jews who lived among the Bedouins throughout the years—whether to study them, be trained by them, or undertake a combination of the two—was small, yet their stories captured the imagination of the Jewish community as a whole. Such was Pessah Bar-Adon (1907–1985) (see figure 3.1), a Polish immigrant at eighteen, who lived with the Bedouins in order to conduct research as part of his Oriental studies program at Hebrew University. In his quest to understand why so many biblical leaders were shepherds, Bar-Adon became trained as a shepherd, received an Arab name (Aziz Afandi), and published several books on his experiences as well as short stories relating to the lives of the people and animals of Palestine. Bar-Adon, who was considered "a little crazy" by the growing shepherds' community, finally deserted shepherding and became a famous archaeologist.

Not all members of Haro'e considered the creation of a Jewish Bedouin tribe their goal, and in reality, most Jewish shepherds in the last days of the Ottoman rule were hired individually by the older private Jewish settlements, the *moshavot*. Furthermore, efforts to create Jewish shepherd tribes largely failed in the 1900s and 1910s, underscoring a larger debate about the role of the Hebrew shepherd in Palestine.[25] The shepherds mostly disagreed about the relation between controlling the sheep and controlling the land. Many of them perceived the practice of shepherding as a way to "guard the land" and they considered Haro'e to be part of the larger movement of the Jewish armed riders, "the guard" (Hashomer).[26] For that reason, these shepherds argued, in addition to the traditional flute and cane, the Hebrew shepherd should carry a gun. Others, however, rejected this connection of shepherding to the militant conquering of the land, contending instead

FIGURE 3.1
Pessah Bar-Adon, a Bedouin Hebrew shepherd. *Source:* Aziz Afandi [Pessah Bar-Adon], *In Desert Tents: Stories* (Tel Aviv: A. Y. Shtibel, 1934).

that while they supported the growth of Jewish settlements in Palestine, shepherding was first and foremost a tool to revive biblical Hebrew life.

THE LURE OF SCIENTIFIC SHEPHERDING

Emerging global forces also shaped the formulation of Hebrew shepherding. World War I brought sweeping political and economic changes to Palestine, thereby shaking the larger Palestinian population. With the emerging support from the new British rule and Jewish organizations worldwide, the Jewish settlement in Palestine began to expand, followed by sequential Arab resistance. As a result of growing tensions, the modern Hebrew "New Man" was redefined in relation to not only the diasporic Jew but the Palestinian Arab too (see chapter 4). Shepherds faced this tension head-on; caught between acknowledging the Bedouins' practical knowledge and criticizing them as hindrances to the growth of the Jewish settlement in Palestine, many shepherds started questioning the value of training with Bedouins and seeking alternative solutions.

Zamir, the son of a wealthy citrus grower, was fascinated by the Arab shepherds from his childhood, and in his quest to become a shepherd, debated the road to shepherding in the mid-1920s. "As many did before me," he recounted, "going to the Bedouins was appealing. But I wondered to myself: might they be wrong and this is not the way? . . . I decided to go to Europe, to get a taste of common agricultural practice and to base my work on professional-educational foundation" (see figure 3.2).[27]

A shift away from the dominance of Bedouin knowledge went hand in hand with new ideas about much-needed changes in the land of Palestine and the role of Western science in bringing about these transformations. Science had now become a tool to prove the holiness of the land, demonstrate its remarkable makeup, and substantiate its sacredness.

With the conclusion of World War I and renewal of financial support for various Jewish projects in Palestine, settlers began to reconsider what Hebrew shepherds should and could be. A Hebrew Bedouin tribe as well as permanent shepherds settlements had proven to be inappropriate ways of re-creating the land of the Bible, for both shepherds and sheep alike.[28] "The dream of a Jewish Bedouin tribe dissolved," said one shepherd. "We understood that we are people from the settlements and that sheep herding must penetrate the Hebrew economy."[29] From now on, settlers agreed, it

FIGURE 3.2

"Profession: Élève Berger; Nationalité: Palestinienne." David Zamir's (previously Pishezner) French identity card, issued for the purpose of his shepherding studies during 1928–1929. *Source:* Kibbutz Merhavia Archives, David Zamir Collection, 1.1b.

was not enough for Hebrew shepherds to simply know how to control the movement of sheep in the open land. Instead, sheep had to be integrated into the changing agricultural economy of Palestine and become part of the emerging mixed-farming economy dominated by Jewish settlers. Gershon Fleischer (1893–1974), a leading figure in the Haro'e group, explained how the vision of Hebrew shepherds changed as they "began establishing a shepherds group anew. This time we wanted to . . . raise sheep in greater numbers, and on more economic and scientific foundations."[30] Competing with the success of dairy cow husbandry (see chapter 4), shepherds had to prove to the Jewish leadership and British government that sheep husbandry was worthwhile.[31] Shepherds argued that "rational" sheep management would demonstrate progress, reestablishing efforts through which biblical practices could succeed and expand.[32]

Starting in the late 1920s, sheep pens were built on many of the kibbutzim, alongside cowsheds, children's houses, and communal dining halls.

Under the kibbutz structure, the management of sheep was different. Sheep now had their own houses; they no longer spent the night next to their owner outside or within the family tent when it stormed.[33] Various kinds of people became involved in the lives of the sheep, from the toddlers who came to see and pet them when the sheep returned back from the pasture, to the older children who learned how to feed and treat them as part of their school education, to the young adults who were trained to raise and breed them in agricultural schools and experimental stations.[34] City dwellers came to witness agricultural ways of life and observe the source of their famed *bryndza* cheese.[35]

More important, though, the lives of sheep changed because settlers thought that sheep management needed to professionalize and become "scientific." It was becoming crucial to collect data on the sheep as well as gather statistics on milk yield, diseases, and feeding regimes. Shepherds began organizing and evaluating this data in a herd book, imitating the practices of the cow growers (see chapter 4). Registration of and insurance on the herds became systematized, and medical problems called for consultation with veterinarians and specialists from the Hebrew University. These experts were also involved in experiments on sheep breeding and feeding within the settlements, just like those conducted by Sulman. "Tedious and scientific work," argued the shepherds, improved the quality of the local breed from "its primitive state in the faltering Arab herd."[36] The shift to intensive agriculture facilitated a concentrated effort to increase milk yield too, and to some extent, enhance the quality of meat and wool (although these were never as critical).[37] The sheep were now milked during particular hours and in rapid pace. Furthermore, the shepherds were instructed to refrain from talking. There was no longer a strong need for those special, wild calls (see figure 3.3).[38]

One of the most significant events in professionalizing the sheep industry was the founding of an organization of shepherds, Hanoked (as noted earlier, another biblical word for "the shepherd") in 1929, later to become the Hebrew Shepherds Association.[39] The organization held annual meetings to deliver scientific lectures as well as exchange information about recent developments in the field of sheep management, breeding experiments, and local inventions. Hanoked also started official training programs for shepherds, determined the appropriate Hebrew terminology for the field, and negotiated the future of sheep rearing with both the British

FIGURE 3.3
A Hebrew shepherd of the settlement: no Bedouin dress and no wild calls. Jacob
Rosner, "Shepherd with Sheep Flock Grazing in the Pasture at Kibbutz Afikim," 1945.
Source: Jewish National Fund Image Collection.

government and Jewish leadership.[40] Members of the association attempted
to participate in the larger global scientific community; many read professional literature, some sent their publications worldwide, and others established connections with sheep growers around the world and experimented
with imported breeds of sheep (although these crossbreeds proved unsuccessful).[41] Members of the organization also read and published extensively
in *Hasade*, the scientific journal of the Jewish agricultural community in
Palestine, and later established their own journal, *Hanoked*.

The organization valued its successes, hoping to demonstrate "that they
[the sheep] play an important role both as a factor in maintaining and

increasing the fertility of the soil and as an alternative source of revenue in the diversified system of farming which is . . . the backbone of Jewish colonization in Palestine."[42] Hanoked did so by presenting the sheep and data in special exhibitions, and writing reports sent to the Jewish leadership and British government (see figure 3.4).[43] These must have proved somewhat convincing as "the [British] government approved of loans for purchasing sheep for 26 settlements" in 1942 and sheep pens gradually became a noticeable part of Jewish agricultural life.[44]

Efforts to demonstrate the importance of shepherding were also targeted at the Jewish community at large. Shepherds thought they had a special role to play, and that the success of professional and scientific sheep management was proof of that role. "In the revolution that is emerging among the Hebrew people we are those who bring back the glory of ancient days," announced Zamir in a lecture to students of the shepherds training course. He explained that shepherding was "the most ancient occupation, which is most rooted in the homeland that we are renewing with our work." According to the shepherds, laypeople had to realize that shepherding was not only rational and good for the economy of Palestine but a way to redeem the land too, since "the only person that walks on the soil most of the days of the year is the shepherd. And his [sic] absence brings an orphaned image to our lands."[45] Numbers were key, but so was knowing the land by foot. It was also significant that the wandering sheep were of a stock that was considered ancient, not crossbred.[46] Sheep and shepherding, in other words, were still essential for demonstrating that the land of Palestine could ultimately become the biblical land.

"AN ELECTRIC SHEARING MACHINE WITH VERSES FROM THE BIBLE": RITUAL TECHNOLOGY

Numbers, exhibitions, and conferences were not the only means for highlighting the importance of sheep and shepherds. "We are not just a professional association, but mainly an association for spreading the idea of sheep rearing," noted a member of Hanoked.[47] The production of shepherd culture became just as intensive as the production of milk and ewes. Various folklore practices and forms of art were married to illustrate the relationship between shepherds, the Bible, and the land of Palestine. A diverse group of artists, many of whom were shepherds too, dealt extensively with the

FIGURE 3.4

"Items of Income and Expenditure for Ewe in a New Flock in Peace Time," a hand-painted poster (100 × 70 cm.) prepared for the first Jewish Shepherds Association's exhibition in the "Children Village" at Ben-Shemen, 1945. The poster, exhibiting the profitability of sheep rearing, was painted by David Alef (Alkind), who was both an artist and shepherd. The numbers and illustrations show that the sheep were raised for their milk; the income distributed shows milk (56.6 percent) as opposed to meat (29.4 percent), manure (9.4 percent), or wool (4.7 percent). *Source:* Hashomer Hatzair Archives, Yad-Ya'ari, Givat Haviva, drawer 94–4.91(1). Alef, who was a shepherd at Beit-Alfa, Ramat Yohanan, and then Beit-Hashita, built his first atelier inside the sheep pen.

practices surrounding sheep husbandry.[48] From the early twentieth century on, prominent painters, photographers, and sculptors used the shepherd, sheep, and goat as the main subjects of their works.[49] Moreover, much like for the Bedouins, the song on the lips was as crucial as the knowing foot on the land; many dozens of poems were written and composed into songs, which were then sang on the holidays in both rural and urban settings.[50] "Know Dear Shepherd" was one such song, written by Matityahu Shelem, a prominent member of the shepherds' community: "Know dear shepherd / that springtime has arrived / Descend from the mountain to the valley / On the pasture you shall expand / Your song will erupt, and will astonish in might / Ancient shepherds' singing.""[51] Geographic expansion and movement are also central to another one of his songs, "The Sheep Have Spread," similarly written and composed by Shelem. Both these songs, as many others, became part of the Hebrew Shearing Holiday (Hag Hagez).[52]

The celebrations of Hag Hagez, which began (or were renewed from biblical times, as the shepherds would say) in the 1930s, were the epitome of the ritualization of shepherding in communal agricultural settlements.[53] Shepherds of several settlements communally sheared the sheep during the day, combining an "electric shearing machine with verses from the Bible," followed by special dances, songs, and plays.[54] The program was rather rigid. As documented by the shepherds, Hag Hagez happened during sunset, when "the public [would] gather by the decorated pen . . . standing on both sides of the road, waiting for the sheep to come back from the field. The shepherds appear with their herds, the sheep are washed and clean. . . . [T]he shepherds come to the center wearing flowers."[55] The rest of the night centered on the sounds of the flute, Hebrew shepherding songs, and theatrical musicals with biblical figures. Most important were the communal dances, which—although choreographed by leading professional dancers— were performed by all participants. Such was the famous "Lamb and Kid" (see figure 3.5).

"Our way is the way of settling the land, making its deserts bloom and fortifying its borders," declared the shepherds in a report from the twenty-fifth anniversary of their association. "We, who wander with the sheep in the paths of the land, in the mountains and valleys, see the blessing that is hidden within it. Indeed a land flowing with milk and honey it is."[56] This meeting, held just a few years after the establishment of the State of Israel, symbolizes the height of the Hebrew shepherds' success. As a result of the

FIGURE 3.5
Illustrated dance annotations for "Lamb and Kid," printed for the shearing holiday celebrations in 1957. The song was written and composed by Matityahu Shelem, who employed Hasidic motifs. *Source:* Hashomer Hatzair Archives, Yad-Ya'ari, Givat Haviva, 125.13, 3.

concentrated state efforts that followed the 1948 war, the number of Israeli sheep pens on both the kibbutzim and moshavim grew substantially. For example, the number of sheep held by Jews grew from nineteen thousand in 1948 to a hundred thousand in 1956.[57] This was due not only to successful breeding but the mass transfer of animals from Palestinian to Jewish hand along with new and emerging government regulations too. Many hundreds of sheep were given or sold cheaply to Jews by the military after their Palestinian owners were forced to leave them behind during the war, or in other cases, after the military confiscated sheep (as well as goats and cows) owned by those who managed to stay.[58] The new government also began sending members of the Hebrew Shepherds Association across the northern borders to buy sheep from countries that were now considered enemies.[59] At the same time, as part of an attempt to decrease the number of sheep outside Jewish settlements, the government started forbidding the purchase of sheep across borders for Palestinian Arabs now under military rule.[60] In addition, in the early 1950s, the Jewish Agency examined the possibility of importing sheep from places such as Australia and Argentina, and during 1953–1956, finally organized the purchase of many thousands of sheep from Turkey.[61] For a while, and with the support of various state institutions, it seemed that sheep had indeed managed to become "one of the pillars of the land's economy."[62]

The celebrations of the twenty-fifth anniversary were not complete, however. In spite of the dramatic growth in the sheep population, Jewish shepherds attending the celebrations were also oftentimes pessimistic, noting, "Bad winds of crisis will not cause despair in our hearts."[63] Indeed, trouble was mounting on all fronts. The move toward intensive sheep management had generated a financial problem: feeding the sheep inside the pen, instead of relying on pastures, made the sheep business too expensive to be worthwhile. The adoption of milking machines as a way to deal with rising costs created its own complications and was ultimately insufficient.[64] There were also problems with selling the milk to the dominant agricultural cooperative, Tnuva, which preferred cow's milk over sheep milk (see chapter 4).[65] Few agriculturalists agreed to work with the sheep that were still around, with complaints that they stank.[66] Finally, the shepherds of the mid-twentieth century said that the spirit and cultural value of raising sheep was lost.[67]

From the late 1950s onward, the various publications of the organization adopted a tragic tone. With each passing year, more and more announcements about the closure of pens reached members of the organization, causing one shepherd to remark, "We shall admit that in mixed feelings these words are written . . . as we strive to the future satisfied with our professional achievements, the harsh feeling that our numbers are declining would not let go. We suffer tremendously from the extermination of the herds, a phenomenon that has no justification."[68] By the late 1960s, sheep management was in a severe crisis, and the smells and sounds of sheep were almost gone from the settlements. Despite their hopes and efforts, Hebrew shepherds had become extinct.

REBELLIOUS SHEEP

Why did the Hebrew shepherding project fail? Bar-Adon, a first-generation Hebrew shepherd who lived with the Bedouins and learned how to manage herds in the open, chose one particular sheep to explain the magnitude of change in sheep management, particularly the sedentarization of life. Bar-Adon crafted a story of a sheep named Shatra/Bat-Hayil (Arabic and Hebrew

for "skillful girl") in a narrative that was first published in a literary magazine in 1933 and later as a book in 1942. In presenting his story to young
readers of Hebrew during a time when a new type of sheep management
was taking form, Bar-Adon sought to critique larger social transformations.

According to the story, Shatra was born and raised in a nomadic tribe,
where she grew up to be a natural leader given her good character and in
fitting with her Arabic name. Her early years were happy ones, and she
appreciated the open space. When she was fully grown, Shatra was separated from her herd and taken against her will to the city market, where
she was sold to a new owner. Now living among unfamiliar sheep in a new
pen and a permanent settlement, Shatra, who now had the Hebrew name
of Bat-Hayil, showed signs of distress. "The shepherd did not know what
was wrong with Shatra," notes the narrator. "She looked healthy, beautiful,
nice and loved by everyone, just as she used to be in her herd in the tribe.
But something strange had happened, and the shepherd could not figure it
out." For the first time in her life, Bat-Hayil was not comfortable wandering: "She walked with the herd but looks gloomy, moving to the sides, as if
she wanted to hide her presence . . . as if she was lonely within the herd. . . .
[O]nly at time she would made a long 'behhhhh' sound. . . . [T]here was
always fear in her eyes, a feeling of insult."[69]

In the last part of the story, Bat-Hayil sits inside the pen at night, remem-
bering "her sunny homeland, her herd, and even her shepherd, that was
mean to her once. . . . [S]he looked at her fellow sheep. . . . How can they
all sit peacefully and quietly?" As her fury grows, Bat-Hayil "shout[s] out
loud, as if she was about to be slaughtered. . . . [W]hat good does pasture,
plenty of water, and rest in a good pen do, if she was locked? Where are the
nights of wandering far-far away . . . where are the mountains, the valleys,
the open space?" The sheep stands up and pushes "to freedom, to spacious-
ness, to the mountains, to the hills, to the valleys, to the sun, moon and the
stars—to the godly space. To the wind and storm. . . . [S]he burst joyfully
outside," and all of a sudden, the rest of the sheep follow. The guard notices
too late, and as the sheep reach the paths of the mountains, they begin to
laugh out loud (see figure 3.6).

Through the story, the reader comes to understand that the appearance
of comfort, order, and plenty in the pen was only a deceit, and the sheep's
ultimate happiness depended on open, wide spaces along with movement
and change. Freed of the scientific, productive, and rational reality inherent

הִיא הִתְפָּרְצָה בְּשִׂמְחַת גִּיל הַחוּצָה

FIGURE 3.6
"She burst joyfully outside [the pen]," Shatra/Bat-Hayil by artist David Alef, in Aziz
Afandi [Pessah Bar-Adon], *Among the Herds of Sheep (From the Stories of a Shepherd)* (Tel
Aviv: Ahi'asaf Publishers, 1942), 95.

in settled sheep rearing, sheep excel in nomadic life. In a period when a new
manner of raising sheep was taking shape, Bar-Adon's tale of one rebellious
sheep bears a strong theme: permanent settlements are no place for sheep.
The moral of the story is possibly greater, relating to the parallel between
the sheep and shepherd: permanent settlements were no place for shepherds
either. According to this perspective, scientific shepherding was an oxymoron.

CASTAWAY

In the late 1950s and 1960s, every communal vote to close a pen was devastating for the larger shepherding community, which tried to battle the
trend with all its might.[70] The lofty vision of Zamir, the main advocate

behind the power of the sheep population to transform the land, had
failed.[71] Zamir's disappointment was significant, contributing to his sense
of personal ruin.[72] Other shepherds, however, even when it was clear that
they had failed, thought they had created something unique:

> We paved our own road [in sheep management]. . . . [T]he attempt to imitate the
> system of the Bedouins in some manner was only a short episode. . . . [From] the
> rest of the world we couldn't learn much either. Sheep farmers in the developed
> countries were [a] model for us as far as the structure of the farm and the value of
> the sheep goes, but not . . . for the work methods and breeding, or the ways of life
> of the shepherd. In that we are better and very different from them.[73]

The need for a unique way of shepherding in Palestine was part of a growing
desire to expand Jewish settlement in Palestine. Even when Arabs were considered the link between modern Palestine and the land of the Bible, European settlers still believed they had the right and ability to settle the land,
and more important, tools to redeem it. And when the Arabs were no longer
seen as the missing link to biblical life, the shepherds continued to consider
the sheep as "the iron bridges that will connect between the past and the
future of our renewing ancient homeland."[74] The sheep of this time were
not crossbred, as shepherds wanted to preserve the belief that their sheep
were Indigenous and had been wandering the land since ancient times.[75]

For the first Hebrew shepherds, smells, sights, and sounds were important ways of knowing the sheep and controlling them in their larger environment. But the use of the senses changed with the shift in shepherding.
Instead of calling and screaming, shepherds were silent during milking and
sang only on holidays; as opposed to moving in space, shepherds danced
near the pen; and rather than observing other shepherds to learn the art
of shepherding, they became the subjects of the artistic gaze themselves
(see figure 3.7). Indeed, the distinct smell of sheep was no longer useful as
it had become yet another reason to close down more pens.

The change in the agricultural economy of Palestine also induced a shift
from extensive to intensive use of the land. The way to expose the biblical
land—by means of agriculture as well as archaeology—was to dig deeper. In
this new method of sheep management, the sheep did not graze but instead
were fed inside the pen. Thus those shepherds caring for the sheep spent
less time herding the sheep, and more time measuring, feeding, and milking them as well as cleaning the pen. In this new and for a time successful
form of sheep management, which involved little herding, there was not

FIGURE 3.7
Scientific shepherding: an oxymoron. Poster for the twenty-seventh annual meeting of the Hebrew Shepherds Association, 1956. *Source:* Hashomer Hatzair Archives, Yad-Ya'ari, Givat Haviva, 125.2, 2.

much room left for Hebrew shepherding. The Hebrew shepherd was therefore in essence a nonscientific figure. Standardizing sheep management entailed getting rid of the shepherd, even though shepherding was the real goal all along. For this reason, Hebrew shepherding was a story of inevitable failure. The "scientific revolution" in sheep management failed because it undermined the main motivation for bringing it about.[76]

Shepherding was not merely about "going native" but rather about becoming native, or better yet, turning Indigenous. For that purpose, it was necessary to hold an ancient stock and wander the lengths and widths of the land. With attempts to shift to "scientific shepherding," holiday dances and the welcoming of sheep home from the pasture remained important, as these were part of the performance of homecoming. Shepherding was always a stinky business and never profitable, yet the question remains, Why did Jewish shepherding last as long as it did? Why did it cease to last when it did? Larger changes in the kibbutz and state economy are a partial explanation. With industrialization and urbanization, a new way of tightening bonds with the land grew stronger.[77] Making a land of plenty was no longer about experiencing the environment, but about successfully manufacturing its products.

Finally, during this period, the majority of those Bedouins who managed to stay with their animals and live on the land despite the war and changing political regimes were gradually forced to abandon shepherding as well. As we saw in the previous chapter, both British rule and the Israeli government restricted the grazing of animals throughout the lands of Palestine/Israel, as the process was considered harmful to afforestation efforts. Furthermore, government efforts sought to establish the sedentarization of Bedouin life.[78] These pressures asphyxiated Bedouin agricultural habits; thus the failure of the Hebrew shepherding project is entangled with the destruction of Bedouin ways of living and shepherding.[79] While most Hebrew shepherds were able to make successful career shifts, Bedouins became constrained and invisible in the eyes of the state.[80]

4 HOLY COW! MILK YIELD AND THE BURDENS OF THE "NEW JEWESS"

The Israeli cow reigns supreme. She has the largest milk yield in the world, on average 10,500 liters a year, compared to 9,500 liters per cow in the United States and some 7,500 liters per cow in Europe.
—Yaron Dror, "Udderly Marvelous Gina: Israel's Most Productive Cow"

IVF [in vitro fertilization] soon became a field of internationally acclaimed Israeli excellence and a source of national pride . . . and Israeli women are the world's most intensive consumers of the technology.
—Daphna Birenboim-Carmeli, "Contested Surrogacy and the Gender Order in Israel"

At the age of two and a half, Stavit started lactating. It was 1935, and after that, she became accustomed to a rhythm three times a day. Other cows in other places grew to produce milk only twice a day, most by hand, but some with the aid of a machine. But just like her, they were forced to be pregnant every year of their adult life in order to ensure the flow of milk. Stavit ("autumnal" in Hebrew) was part of an early generation of crossbreeds—between a Dutch bull and Syrian cow—that formed the foundation of the settler dairy industry in Palestine (see figure 4.1).[1] From an early age, Stavit was "flowing with milk, but she required special care, because it became clear that only a permanent and experienced milkman [could] maintain the stability of her milk yield."[2] She excelled in managing these daily and annual tempos for an exceptional number of years, and gained fame in 1950 at age seventeen when it was realized that she produced over a hundred thousand

FIGURE 4.1

Stavit, an exceptional Hebrew cow. *Source:* Aaron Ever-Hadani, *Agriculture and Settlement in Israel: A Decade after Its Establishment* (Beit Dagan: Ministry of Agriculture, 1958), 141.

liters of milk during her lifetime.[3] Stavit then received an exceptional prize for her outstanding labor, and was allowed to retire and live as an unproductive creature until she perished as a result of the unusual cause of "old age."[4] Her tenders noted that she was particularly remarkable as she was aging: "Her body is healthy and strong, her udders are healthy . . . her teeth . . . appetite and ability to utilize food are admirable. Her fertility is flawless: fifteen births in fifteen years (she even had twins once)."[5] Soon after, the so-called Israeli cow was proclaimed a global milk-yield champion and became a leader in the consumption of artificial reproduction technologies.[6]

The Christians and Jews who traveled or immigrated to Palestine considered themselves to be modern and scientific; they also thought of Palestine as the land of the Bible and expected it would be fertile. They chose science and technology as appropriate means for redeeming the land. In this process of creating plenty, settlers concentrated many of their efforts into milk production. Since the early years of European settlement in Palestine, researchers in dairy farming have taken pride in their milking and reproduction technologies, and celebrated their record highs in increased milk yield.[7] The dairy industry, milk, and cattle management have been fundamental elements in the economic, political, ecological, and cultural history of Palestine/Israel and Zionism, yet they escaped scholarly attention.[8]

Milk has proven a good liquid with which to think about industrialization, urbanization, modernization, capitalism, colonialism, and nationalism.[9] Environmental historians have used cows to demonstrate the relation between farming practices and social and economic orders.[10] Recent studies also investigate changes in cattle husbandry and breeding practices, and yet others examine the connections between aspiring for enhanced milk yield and managing cow reproduction.[11] This chapter follows this body of work and explores the conditions that supported the emergence of a particular kind of breed—the Hebrew cow—in Palestine in the first half of the twentieth century. A story of success in one sense, the development of a dairy industry is a tale of struggle and failure too, as this resource-thirsty industry made little economic and environmental sense in its time, and as cows' bodies were challenged by heat and infertility.

This chapter focuses on the already-established Christian German Templar settler community and fast-growing Jewish settler community under British rule in Palestine (1917–1948), and looks at the efforts that were put into making Palestine a land of plenty by producing a lot of milk from many female cows. In addition to several champion cows, three male figures are the protagonists of this story. Yitzhak Elazari-Volcani (formally Wilkansky) (1880–1955) and Yehoshua Brandstetter (1891–1975) both promoted the idea of the Hebrew cow and its centrality to European settlement in British Palestine; Efraim Smaragd (1902–1976) advanced similar ideas with his designing of a milk-yield champion breed in the late British rule in Palestine and early years of the State of Israel, despite environmental and economic pressures. Beyond a discussion about female cows and male designers, this chapter illustrates how in the minds of settlers at large, dairy cows gradually occupied a central position in the efforts to change the land of Palestine and make it plentiful.[12]

Along with milk production, animal and human reproduction has played a central and symbolic role in the creation of plenty. Achievements in fertility research have been celebrated since the early decades of the twentieth century, and in recent years, much like cows, Israeli women have become global leaders in the consumption of fertility treatments.[13] The final part of this chapter examines the focus on the fertility of the newly invented Hebrew cow, connections made between milk production and reproduction, and ways in which the fertility of bovines and humans alike conveyed special providence.

TERRA MOOLLIUS

> There were days when the immigrants were complaining that the land "flowing with milk and honey" was only true in the Bible and not in reality. The guests at the hotels and the hospices would always hear the "traditional" answer from their hosts: there is no milk and no butter in the country; and when they would return to their county and to their homeland they would tell about the poverty and crowdedness of the land of the fathers, about the curse that still lays on it and about bitter life etc., etc. That has changed in the past few years: dairies have been established in the country and throughout the year milkmen of OUR BROTHERS are supplying fresh milk and butter and cheese to their customers, and just as in the past consumers were complaining about the lack of milk—now the producers are complaining about the lack of market.[14]

Cows in Palestine labored and were bred for a variety of purposes, and their bodies supplied more than just milk. Palestinian Arabs usually raised cows in small numbers and commonly used them as beasts of burden.[15] Cows' manure enriched the soil as they grazed, and their flesh became food when they died. Much like honey, cow milk and the more prevalent goat milk were mostly limited to seasonal and local uses. Despite the fact that they were more common to the land, goats and their milk came to be depicted negatively by settlers and the changing governing rules.[16] Europeans seemed particularly disappointed by the small amounts of cows and cow milk they found: "Cows have always been in disadvantage. In a country without grass, 'deep uddered kine' are not even thinkable.[17] . . . [I]n the land that is said in some mysterious way to have 'flowed with milk,' nowhere with the friendly cow."[18]

As part of the larger British imperial plans for improvement and development, milk gradually became a pillar substance for the agricultural economy. It was also advocated as a basic, yearlong food staple for all populations in Palestine, settler and Indigenous alike.[19] Dairy cows also fit well with European ideas about agricultural settlements.[20] Just as settlers from Europe brought cows to the Americas, settlers from Europe brought cows to the Middle East.[21] Yet the idea that Palestine was a unique place that should prove bountiful (again) added a crucial dimension to the move to make intensive use of the land and efforts to create a significant dairy industry. The extent of these attempts becomes obvious in light of the environmental conditions the settlers faced. The Palestinian landscape and climate at the turn of the twentieth century was distinct from that of northern and western Europe, and thus intensive cattle raising in Palestine quickly became a struggle. Nevertheless, cattle husbandry and milk production became cardinal to European governance and settlement in Palestine, particularly to the expanding Jewish settling population.

The first efforts to intensify cow milk production in Palestine are usually attributed to the German Templar society that settled in Palestine in the second half of the nineteenth century in order to prepare the land for Jesus's second coming. The settlements established by this independent group, the pietistic outcasts of the Lutheran Church in the state of Württemberg, were the first to survive among various failed projects in Palestine during that period, such as the agricultural farm in the village of Artas (see chapter 1) and American Colony near Jaffa.[22] Beyond their efforts to better the land and their hold over it, the Templars took part in the project of the scientific deciphering of Palestine.[23] Their farms, which depended on Arab labor, were based on the mixed-farming model they were familiar with in Germany, and they considered dairy cattle the supporting pillar of the farm—supplying foods for farmers and for sale as well as manure for farm crops.[24] As part of their attempt to launch the dairy farming industry in Palestine, German practices included importing small numbers of European cattle (mainly bulls) to Palestine as well as initiating crossbreeding with several kinds of local cows.[25]

The success of the Templar agricultural settlements was important to many Europeans—both Christian and Jewish—as these settlements were perceived as spaces for experimentation that produced knowledge about the potential for land revival in Palestine.[26] Jewish settlers studied and adopted many of the German farming and settlement practices, and as these settlers' attempts grew more successful, they began to raise dairy cattle on Jewish experimental stations and farms.[27]

During the first decades of Christian and Jewish cattle husbandry, the number of cows managed as well as the amount of milk produced grew rapidly and dramatically.[28] But it was only in the 1930s that cows became really central to European governance and settlement efforts. As the numbers of imported and bred cows increased, and farmers moved from extensive to intensive use of the land, the new dairy industry came to center on cow milk. The 1920s and 1930s are particularly important to the creation of organized cattle management for the purpose of milk production, and consequently included the establishment of several agricultural professional organizations and their journals.[29] Other new practices of this time involved controlled investigations at experimental stations and on some of the settlements; the documentation, collection, and comparison of statistical data; and the employment of the first veterinary doctors. Jewish domination over the cow milk market became evident in the early 1930s vis-à-vis shrinking Templar settlements in the face of mounting British-German tensions along with growing friction between German and Jewish settlers. Indeed, the industry became one of the largest and most profitable agricultural industries—second only to the citrus industry, which produced the famous "Jaffa" oranges.[30] The earlier focus on selective breeding gave way to experimentation with artificial insemination, which received wide support in the 1940s.[31] Additionally, mechanization of milking as well as the growing sophistication of milking technologies emerged and expanded in these formative years, if not with some pushback from farmers.[32]

Beyond the industrialization of milk production, and as opposed to other profitable agricultural industries like citrus, the dairy industry grew exponentially in spite of rather low dairy product export and, in the eyes of the producers, low demand.[33] Furthermore, and until recently, the breadth of the dairy industry was significantly greater in Israel than in neighboring countries.[34] From an ecological point of view, the growth of milk production in British Palestine and the State of Israel has been remarkable. From the early days, dairy farmers and researchers complained about the problems created with intensive cattle management in the Palestinian environment. Yet these farmers persisted in their efforts to increase milk yield, despite managing the Middle Eastern climate along with its effect on bovine production and fertility.[35]

Milk and dairy have also occupied a central cultural role for Jews in Palestine and the emerging Israeli society. Most Israeli Jews enjoy dairy, consuming a large variety of soft cheeses, and often celebrating the nutritious and national value of milk.[36] Somewhat similar to other national programs, the growth of the dairy industry might even be viewed as a symbol for the success of the Zionist project.[37] The celebrations of Shavu'ot, a harvest

FIGURE 4.2
Poster made for the celebration of Passover in the cowshed of Kibbutz Mizra' (1947) depicting a vision for the ideal socialist life in plentiful Palestine. It reads, "Our heads will be filled and filled with dew. The blessing of seed will not betray." The word "seed" in Hebrew (*zera*) also means sperm, thereby the sentence might be read as "the blessing of the sperm will not betray." *Source: Toshek: Avraham Amarent* (Tel Aviv: Sifryiat Poalim, 1993).[39]

holiday, are one good example. This holiday is dedicated to the land, but specifically centers on milk and dairy-based cuisine, accompanied by white clothing. Interestingly, several of the communal settlements have even celebrated the Passover holiday inside cowsheds (see figure 4.2), often the first building constructed by these communities.[38]

"THIS LITTLE LAND HOLDS THE POWER TO SUPPORT A MILLION COWS": VISIONS OF PLENTY

The growth of milk production and success of a Jewish-dominated dairy industry are deeply associated with the work of a particular agronomist, botanist, and writer: Yitzhak Elazari-Volcani. This Lithuanian Jewish settler immigrated to British Palestine in 1908, established several (and some of the first) agricultural experimental stations, and later founded an institute for agricultural research, which remains the central such institution to this day and is now named in his honor as the Volcani Institute. Elazari-Volcani was a prominent figure in the design of the agricultural-economic approach (the mixed-farming model) and Jewish semicommunal agricultural settlements (the *moshav ovdim*) as well as the promotion of milk production as early as the 1910s.[40] In particular, this "mentor for the settlement of the people on their land" or "architect of the settlement" was advocating for a shift in agricultural practices in Palestine.[41] As the writing of and about Elazari-Volcani demonstrates, the settlers' imagination of the biblical landscape, scientific agriculture, and production of milk were entwined in early twentieth-century Palestine. It also illustrates how this process relied on Palestinian Arab agricultural knowledge, and how the appropriation of that knowledge was gradually denied.

Elazari-Volcani was among the first generation of theorists and public figures who contributed to the economic and agricultural dominance of Jewish settlers in early twentieth-century Palestine. He was a prolific essayist too, and published under the pseudonyms of A. Tsioni (literally "A. Zionist") and Ben-Abuya (the heretic Tanna of the Talmud).[42] Throughout the years, Elazari-Volcani explained his vision for the development of the agricultural economy in Palestine, and among Zionist thinkers of the period who debated the nature of the Jewish settlement in Palestine, Elazari-Volcani was considered *the* rational man. As opposed to the "dreamers" or those who just wanted to reconnect with nature, noted one scientist, he was "carrying the flag of the rules of economics, his whole world view is rationalistic-materialistic. He denies that the spirit has the power to influence life."[43] It is precisely because of this reputation that an analysis of his writing is illuminating; his various publications show that he considered the land of Palestine to have special qualities that had nevertheless been suppressed and regarded modern science as a tool for revealing them. Elazari-Volcani's publications also explain why he thought milk should become the center of the agricultural economy of Palestine and the justifications he found for the centrality of raising cattle there.

In his writings, Elazari-Volcani described the land of Palestine as European settlers from the late nineteenth century commonly did, writing of a land that possessed special powers, which settlers are supposed to redeem.[44] In *The Design of the Agriculture of the Land (of Israel)* from 1937 he wrote that "there are few lands where the gap between the possibilities and actuality is so great as in the land of Israel. Enormous are the powers that are hidden within the land, and are currently narrowed to the surface of the soil."[45] As he continued in this book, Elazari-Volcani laid out his vision for the future of the land; specifically, he discussed the depiction of this future in the Bible, explained the manner in which this vision could become a reality, and rationalized who was equipped to hear such secrets of the land:

This is the vision of population density and the continuity of endless crops, with no stop. The prophet of population density was "the shepherd, and the one taking care of sycamore-fig trees"—an expert in agriculture.[46] His vision shall become a reality, not with magic but with course of nature. The wisdom of man, the power of his vigor, the fire of his belief, the persistence of his love, the wealth of his strength and fortune—these are the ones that will change the face of the earth and will transform the desert into a garden of beauty.[47]

The one capable of enabling bounty and therefore redemption in Palestine is none other than the scientist, the expert in modern agriculture. With the knowledge of scientific agriculture, Elazari-Volcani argued, the biblical vision of a land of plenty could become a reality.

The way to make this biblical vision a reality was by increasing the number of cows. "'The land flowing with milk and honey' might seem to many an exaggerated phrase," he noted, "if because the people of the east exaggerate or because of the minimal achievement of those lost in the desert. But the prophesy of the day will become the grey reality of tomorrow. This little land holds the power to support a million cows instead of the sixty thousand that it does today."[48]

One of Elazari-Volcani's projects, which he published under the title *The Fellah's Farm* in 1930, centered on the attempt to quantify Indigenous Arab agricultural knowledge and practices, with the stated goal of testing the benefits of the mixed-farming model in Palestine.[49] Making agriculture quantifiable was only one way to illustrate the importance of experts' scientific solutions. But this did not mean that science alone was sufficient. Prior to the 1937 publication, Elazari-Volcani had long warned about the dangers of modern technology and mechanization.[50] In *On the Road*, for example, published in 1918, he highlighted how "European machines" alone were not the solution to redeeming the land, just as the fellah's way of working the land was not good enough. He argued instead that the combination of working the land of Palestine with the appropriate application of scientific knowledge would bring redemption.[51]

In his 1912 essay "On the State of Farming," Elazari-Volcani had already analyzed the problems with Jewish settlements and ways in which the raising of cattle would solve them. He lamented that cattle raising, although so common in biblical representations and "the crowning glory of farming," was "completely absent in our peasants' economy." Thus Elazari-Volcani suggested that the creation of this protoecological balance through the raising of cattle would require an entire overhauling of the agricultural system. Ultimately, this dairy-cattle-based agriculture would enable "the revival of a dying people."[52]

As such, Elazari-Volcani was advocating for this change in the agricultural economy of Palestine as a means to create plenty. And after much debate about the model of agricultural settlement throughout the 1920s, the ideas of Elazari-Volcani and his circle had a critical influence on shaping this economy.[53] Some settlers argued that the combination of religion and science in Elazari-Volcani's theories was what made their application so successful.[54] Others, however, considered the focus on dairy farming as the basis of Jewish settlement in Palestine to be "rational" and independent from religious or romantic ideals.[55] This assertion, along with the wide acceptance of Elazari-Volcani's view of dairy farming as an economically and environmentally preferable solution, are particularly striking in light of its costs; to be sure, the production of milk in British Palestine and the State of Israel has been considered highly successful, but environmental aspects have made it extremely difficult and expensive too. These troubles varied, and included acclimatization, heat, disease, and the availability of only limited grazing areas. While the success of the industry was great—so much so that "the cowshed [wa]s drowning in surplus of milk," as described by a member of the Jewish Cattle Breeders Association—the costs were extremely high. Levi Eshkol, at the time Israel's agricultural and development minister, then finance minister and later the prime minister, and widely known for his support for farmers, also famously declared in the 1950s that "the cow is the number one enemy of the economy of the state of Israel."[56]

INVENTING A HEBREW COW

Colonial powers and European settlers (both Christian and Jews) cooperated in promoting a particular model of agriculture in transforming Palestine. In the 1920s, the establishment and management of the Jewish experimental stations—with the official encouragement of the British government, financial support of Jewish organizations, and technical support of the German Templars—gave Elazari-Volcani and his colleagues the power to orchestrate such changes in the structure of the agricultural economy.[57] While these stations and the governmental station at Acre were built for

FIGURE 4.3
Milking training at the Hannah Meisel Agricultural School for Women at Nahalal and a hammer, 1936. *Source:* Israel State Archives, Zoltan Kluger Image Collection, TS-3/10400.

the purpose of acting as centers of scientific knowledge and authority for all agricultural work in Palestine, they predominantly benefited Jewish settlers.[58] Jewish experimental stations and consequently Jewish training programs and settlements followed Elazari-Volcani's focus on milk production (see figure 4.3). Specifically, settlers attempted to create an appropriate kind of cow for these transformative changes—the so-called Hebrew cow.

One enthusiastic promoter of the Hebrew cow project, albeit for a brief period, was a Jewish settler named Yehoshua Brandstetter (1891–1985). Two decades before becoming one of the first filmmakers in Palestine and the manager of Habima (what would become the national theater), Brandstetter was enchanted by the dream of raising dairy cows in Palestine. On his arrival from Poland to Palestine at age eighteen, Brandstetter described his emotions of awe: "It was only a few days ago that I felt the dirty snow of the [European] town, and here I am now walking on the foot of the

Carmel mount. . . . [T]hroughout the hike I mumbled 'flowing with milk and honey,' and at times, walked away from the route to the side of the mountain, believing my eyes would spot honey between the cracks."[59] Life in Palestine—full of new loves and challenges, many involving cows— would soon take shape for Brandsetter.

His love for cows began on his family farm in the Jewish settlement of Yavne'el in the early 1910s.[60] Given the adoption of the mixed-farming model that had proven successful among the German Templar settlements, few cows actually lived among other animals on the farm, yet according to his biographer, Brandstetter had a special fondness for cows as well as the ladies. In that period, Brandstetter started to advocate for positioning high-yielding dairy cows at the center of the agricultural economy in Palestine. His journey with cows included dairy farming studies in the Netherlands after the Great War, work as a dairy cattle instructor on various Jewish set-tlements in the early 1920s, and the importation of Dutch bulls throughout the 1920s. At the heart of his early work, however, were his continuous efforts to establish the Institute for the Breeding of Dairy Cattle in the Land of Israel.[61]

The idea of establishing an institute that would focus on developing a high-yielding dairy cow won the general support of several Jewish institu-tions in Palestine and the financial support of the Jewish community in the Netherlands. But not everyone agreed that cows were that important; several settlers argued that it would be both a financial as well as structural mistake to invest so much attention in cows at the expense of other parts of the growing agricultural economy.[62] These fierce objections—highlighting that the choice of milk was far from inevitable—finally caused Brandstet-ter to abandon his dreams for an institute. In turn, he adopted new (artis-tic) routes to promote ideas of revival. In 1933, Brandstetter and his wife together produced the *Land of Promise*, a propaganda film depicting various aspects of life in Palestine, and winning acclaim in the *New York Times* as an "excellently photographed and skillfully edited record of the rebuilding of the Jewish homeland in Palestine."[63] "With modern machinery," says the film's voice-over, "the Jews bring back to Palestine its long neglected fruitfulness."[64] While Brandstetter chose new techniques to demonstrate the significance of reviving the land of Palestine, the ideas behind an insti-tute for the Hebrew cow did not die. In 1932, Elazari-Volcani finally estab-lished the institute on the Jewish Agency's experimental farm in Rehovot.

In spite of a debate regarding the nature of the Hebrew cow, there came to be widespread acceptance by settlers that the local Arab cow was not yielding enough milk.

Settlers' attitudes toward local cows aligned with their views toward Palestinian Arab peasants. In the first decades of the twentieth century, when Europeans saw a connection between the people of Palestine and land of the Bible, settlers studied, and to some extent adopted and appropriated, Arab ways of life and knowledge. By following local practices, Jewish settlers hoped to be included in the connection between the people of Palestine and Hebrews of the Bible.[65] Some settlers acknowledged learning how to rear and milk cows (and other animals) from Arabs, and as we saw in chapter 3, some chose to live with Bedouins for extended periods of training.[66] As opposed to most Christian settlers and the first waves of Jewish immigration to Palestine at the end of the nineteenth century, Jewish settlers in these years tried to distance themselves from European ways of living. As part of the wide adoption of eugenic ideas among Zionists, Jewish settlers despised the image of the sickly, poor, degenerating eastern European Jew and aspired to create in Palestine a new kind of Jewish body—one that was healthy and connected to his land (and always a male): a New Jew.[67]

Early twentieth-century settlers tried to create this new-ancient person and body. As time went by, with the deepening political, economic, and environmental settler colonial intervention in the land, things began to change. Along with the intensifying Palestinian acts of resistance to these changes—which reached their height in revolts of the late 1930s—the ideals of nativity were shaped anew. The Arab was no longer considered the link to the people of the Bible, and was increasingly seen as primitive, ignorant, lazy, and violent. The New Jew from this point was redefined in relation to not only the diasporic Jew but also the Palestinian Arab. Not unlike other contexts of growing Western dominance, this change in the understanding of the people of Palestine went hand in hand with new ideas about much-needed changes in the land of Palestine and the role of science in fostering this change.[68]

This new ideal of a healthy and productive Hebrew body, which was defined in relation to both types of old (that is, the eastern European Jew and the Arab), was applied to designing the bodies of cows too. Attempts to enhance milk production among Jewish settlers, which became organized in the 1920s, were based on the comparison of local "Baladi" or "Jabali" cows to ones from "abroad" (either to other regional cows or European ones).[69] Local cows were considered to be immune to disease, but also rebellious, ugly, and producing only little milk.[70] Syrian cows, in comparison, were taller and better milk producers, and therefore imported to several farms.[71] Similar to German Christian settlers, Jewish settlers admired European pedigree cows and brought several of them to Palestine, mainly from the Netherlands. Europeans in Palestine appreciated these Dutch cows, which produced much more milk than both local and Syrian cows, although these Dutch breeds also got quite sick as soon as they reached the farm. Extensive experimentation with crossbreeding thus became a widespread practice by the late 1920s in order to secure the immunity and survivability of local cows in the environment of Palestine as well as maintain the prolificacy of foreign cows. The British government officially supported crossbreeding with local cows on Arab farms as well and offered the services of Lebanese and a few British bulls.[72]

Various accounts point to the success of these breeding practices in late 1920s and early 1930s. Dora Bader (1896–1996), a cow farmer in the mid-1930s, wrote in her diary, "This was how a new land-of-Israel race was created, one that was immune to disease and the damage of climate. With time it reached the yield of a Dutch cow and even surpassed it, and is considered one of the best races in the world."[73] Jewish settlers were rather pleased by the crossbreeds and proud to declare that they had no "Arab cows" on their farms by the early 1930s. They also argued that the temper and character of the Arab and Syrian cows, "(which affects the practice of milking and probably the production of milk as well) . . . was significantly improving" with the addition of European blood.[74] According to most estimates, the number of cows owned by Jewish settlers tripled between 1922 and 1936, and then doubled again in the next decade (from approximately six to then eighteen thousand, and finally thirty-four thousand).[75] By 1937, 80 percent of all cows in Palestine were crossbred, and in spite of several outbreaks of epidemics and frequent financial setbacks, the milk yield per cow almost tripled in this period.[76]

The focus on improving the milk yield of the cow population was consistent from early on, well before that of the Dutch dairy culture, the model for dairy farming in Palestine and the birthplace of the fathers of its champion cows.[77] Settlers' way of valuing animals differed from European ways, with an overall stress on quantity over quality. As opposed to the British,

Dutch, and German traditions, cows in Palestine were mainly judged by their yield not their external features. Cattle exhibitions were far and few between, and the participants in such cattle contests were almost always the highest-yielding female cows, not the prettiest bulls.[78] In 1937, for instance, the cow Zkufa ("upright") of Kinneret, Poria ("fertile") of Kibbutz Geva, and Haviva ("pleasant") of Kibbutz Ein-Harod became widely recognized for their achievements.[79] Efraim Smaragd, the leader of the dairy cattle community for four decades, a major figure in developing a high-yielding cow, and a character often described as "saturated with love for the cowshed," acted as the judge for these yielding, not beauty, competitions.[80] While Smaragd was known, much like his Dutch trainers, to have "an excellent eye" for cows, he invested more attention in his appreciation for production. His eye for beauty, noted his colleague Israel Palvitch, was reserved to life outside the shed. "He liked looking at beautiful women, expressing his opinion and giving comments," Palvitch said, "and would say things like 'look at that pretty primaparous cow coming from the pasture.'"[81]

Smaragd, who immigrated with a Dutch herd to Palestine in 1924 on graduating from dairy farming studies in the Netherlands, became the secretary of the Cattle Breeders Association, a position he held from 1928 to 1967.[82] Under his management, the association began to shape the daily practices of dairy farming; it initiated and financed the translation of professional literature to Hebrew, established the journal of the breeders association, and organized various courses for Jewish settlers.[83] One of the main projects, which shaped the daily management of cows from 1933 onward and won the financial support of the British government, was the use of the herdbook, a technology of documentation to detail a cow's yields, births, and health problems on a daily basis (see figure 4.4).[84] As indicated by the presence of herdbooks across Jewish communal agricultural settlements' archives, farmers adopted this form of data collection and incorporated it into the rhythms of their work.[85]

The success of the Hebrew cow was acknowledged on an imperial scale. As reports in the Palestinian Arab press demonstrate, imperial agents from Iraq, India, and Ceylon traveled to British Palestine with the hope of purchasing dairy cows as the basis for establishing a dairy industry back home. One Indian colonial farm manager was drawn to Palestine after reading about the crossbreed that Jewish settlers succeeded in maintaining; Palestine's geographic location and climate, "between cold Europe and the very

FIGURE 4.4

"The Committee for Establishing the Herdbook in the Cattle Breeders Association," Kibbutz Beit-Zera, August 1933. *Right*, Efraim Smaragd, Menahem Sturmann, and Uriel Levi. *Source:* Estate of Efraim Smaragd.

hot India," he contended, made it the perfect acclimatization station for the cows.[86] A representative of the Iraqi Ministry of Agriculture was directed by North American agriculturists to the Jewish Cattle Breeders Association and arrived in Palestine in 1941 to buy cattle with proved herdbook registry. He traveled back with six animals along with identifying documents that included their individual pictures.[87]

This growing success was accompanied by great anxieties, however. Beginning in the late 1920s, various farms reported on a decrease in the quality of milk production along with infertility issues among third and fourth generations of crossbreeds. The granddaughters of European fathers and local mothers were becoming weaker and sicker, and producers of thin milk. Some saw this as a sign of the danger of interbreeding and mixing, just as has happened in other colonial interactions, when intimate relations threaten to destabilize power structures.[88] In the midst of the Arab revolts of the late 1930s, the Jewish Veterinary Association declared that new evidence proved that Arab cows were not immune to disease after all and thus crossing with Arab cows would not benefit European cows.[89] In 1938, the

Jewish Breeders Association held a conference on this "degeneration of the breeds," and during it, the debate over the nature of the Hebrew cow was reversed: In the early 1920s, bovine experts had decided to use backcrossing with locals so as to ensure the survivability of the breed; but now in 1938, experts established that backcrossing instead with European bulls was required to secure European traits and gradually diminish local ones.[90]

Despite these changing frameworks, concerns did not evaporate. The problem of decrease in milk fat was worrying, for example, although it was later accepted as necessary for the growth in milk yield.[91] Health problems and the issues of infertility in particular, though, continued to be threatening.[92] Much thought was given to the influence of food on fertility. Farmers needed to provide the right type of feed in order to support "as many as a million cows" in the changing agricultural economy, as suggested by Elazari-Volcani; hence they conducted various experiments to identify such feed. Most specifically, Jewish researchers argued against the use of "external" foods: "Feeding with purchased foods from unknown source," they said, "causes diseases . . . and affects badly on the impregnation of the cow and its yield."[93] Ultimately, as a result of these experiments—both growing crops and feeding them to cows—Jewish researchers decided that locally grown Jewish food was crucial for cow health. Coinciding with great efforts to increase (mainly urbanite) reliance on local settler products in the 1930s—also known as the Product of the Land (*Totseret Ha'aretz*) project—cows were to eat "Hebrew products" so they could be fertile and healthy.[94] These Hebrew cows, with their marvelous Hebrew names, produced "Hebrew milk" for Hebrew consumers.[95]

INFERTILITY PUZZLES AND INVASIVE SOLUTIONS

> Since intensive cattle raising began in this country, the cattle-holding farmer suffers not only from infectious abortions disease (*Abortus Infectiosus Bang*), but also from the frequent cases of infertility. . . . [A]s usual, there is no need looking for the causes for infertility in bulls, but only in female cows.[96]

The problem of infertility and miscarriages in dairy cattle became widely discussed in the 1930s, layered by the debate over issues of crossbreeding. The combination of increased production of milk and the climate in Palestine made it difficult for cows to become pregnant. Naturally, since annual pregnancies were crucial in order for cows to continue producing milk, this concern received much attention by both Jewish settlers and British authorities.[97]

Settlers discussed this problem among themselves and in professional journals. Those working with cows at Kibbutz Mizra' noted "the state of impregnation has not been so satisfying. Out of 125 primaparous cows we had 20 cows that were inseminated more than three times and four that were not impregnated at all and then taken out of the cowshed. . . . The state of impregnation among our primaparous cows has not been good for several years, and we have not received an explanation for that."[98] Although settlers voiced their concerns regarding the fertility of cows and sought the help of experts, their questions remained unanswered: "It is well known how hard infertility disease strikes in the last few years in many herds. Not all the reasons of this phenomenon are known and understood among cow holders, and even the doctors and men of science that deal with this matter stand helpless without knowing how to explain the 'stubborn' infertility."[99] Puzzlement over the relation between the local environment and the fertility of cows and hence the production of milk persisted, creating new problems with proximity and a need for novel ways of managing cows.

"As all members [of the kibbutz] know," wrote a member of Kibbutz Degania-Aleph in 1938,

> impregnation in our cowshed has been very difficult lately, and the crisis of cows who did not succeed in getting pregnant of all sorts of reasons or with no apparent reason—is great. Well, we decided to arrange an experiment with collaboration with [neighboring Kibbutz] Degania-Bet and the Hakla'it [the Jewish Association for Veterinarian Services and Livestock Insurance]. . . . [I]t is very important to us that there is a doctor that sees the cows every day, treats them, and gradually prepares them for normal impregnation. . . . [I]t has been three weeks now that Dr. Shapira treats our cows with artificial insemination.[100]

Settlers had begun experimenting with cattle artificial insemination starting in the early 1930s, after one experienced inseminator and several books arrived from the USSR and gave settlers ideas about increasing chances of impregnation.[101]

Many were skeptical about artificial insemination, but some settlers were excited about the potential of this technique. Uriel Levi of the Cattle Breeders Association took literary liberty in describing one of the first experiments with insemination that took place at the Gvat settlement, notwithstanding the violent connotations:

In a tin roofed scalding cowshed stood the awaiting cows, about six of them. From the bull's yard three men marched assured. One carried a 3-liter glass enema and a long tube; the second "holding hygiene in his hand": soap, Lysol, a towel; and the third—just curious. After getting rid of the filth, washing hands and whispering—either swearing or for relief of excitement—the rubber tube was shoved deeply into the Cow's vagina, and with the order "release" there flew the salt water with the sperm that was squeezed from a pinch of cotton wool. . . . [W]as the cow impregnated? Three weeks later it was determined: No.[102]

Disappointments and changing materials were part of this period of experimentation, but in few years, experts began to perceive artificial insemination as a real possibility. British T. Bell, assistant manager of the governmental farm at Acre, reported on artificial insemination practices in Palestine to the British scientific community in 1938, differentiating between Arab stagnation and Jewish progress:

> The two main systems of agriculture in present-day Palestine are remarkable for their extreme diversity. On the one hand is the Arab peasant, whose agricultural practices have remained essentially unchanged for twelve centuries; and on the other the Jewish colonist, whose methods are the quintessence of modernism. Strange as it may seem, these two diverse cultures have something in common, for both have contributed to the development of the animal-breeding technique known as artificial insemination. In very early times the Arabs are known to have applied it in the breeding of their horses, whilst the Jewish settlers of to-day are developing the same method along modern lines with a view to organizing its application as a routine practice in the breeding of dairy cattle.

Bell described a growing interest in artificial insemination among Jewish settlers, arguing that it "had proved to be effective in combating sterility and also in checking the spread of infectious vaginitis, for no fresh cases of the disease occurred."[103] In times when infertility was threatening, artificial insemination offered new ways to overcome unexplained failures of pregnancy as well as common forms of infections.

Perhaps due to the height of the threat of infertility and disease, artificial insemination became a common practice a few years earlier in Palestine in comparison to western Europe and North America.[104] It was widely practiced by the late 1930s, standardized in 1939 as the community of veterinarian doctors began to meet to discuss and study the procedure, and normalized as a course of study in 1946 at the Rehovot Experimental Station.[105]

Nevertheless, in order for artificial insemination to gain prominence, settlers had to finds ways to deal with issues of proximity, as the new fertility practices involved unique management of the distance between cows and bulls. Smaragd, the Cattle Breeders Association's secretary, wrote to the Soviet Institute for Artificial Impregnation near Moscow to ask for professional advice, hoping to send "a veterinarian, Palestinian subject . . . in order that he might perfect himself in this work . . . and the system of transferring the sperm to greater distances."[106] The reliance on the transport of bull sperm created new spatial problems in times of changing political climates. The semen of bulls was first transported on land by bicycle, cart, and bus as well as by boat across the Sea of Galilee.[107] During the 1948 war and when the roads were blocked, the sperm of bulls supposedly reached dairy cows in Jewish farms by air with the help of carrier pigeons and the guidance of one British officer.[108] The flight of semen was not restricted to the Palestinian skies; just as the holy queen bee had been sent by mail from Ottoman Palestine to the United States (see chapter 1), Smaragd flew a famous bull's sperm from the United States to Palestine in 1947. The break of war, however, interfered with records of those cows inseminated by Si'on (meaning "achiever" "record breaker"), a bull whose Hebrew name attested to his record, stamina, and sexual potency.[109] With improving techniques, and a dramatic reduction in the proximity of cows and bulls, artificial insemination finally replaced older forms of breeding. The Hebrew cow, whose creation was now made possible with various means of transportation and the experienced hand of the human inseminator (at least one of whom later became an archivist; see figure 4.5), never had to meet bulls again.

MANY BABIES AND MUCH MILK

Cows' fertility continued to occupy the minds of farmers and researchers.[110] One journalist who followed a veterinarian doctor of the Hakla'it, the Jewish Veterinary and Livestock Insurance Services, offered some indication of the attention paid to the fertility of cows: "From what we have seen in our travels with Dr. Eilenbogen, most of the work of the veterinarians of the Hakla'it is treating cows and their 'sex problems.'"[111] Infertility not only jeopardized the efforts to increase milk production but also threatened the very creation of a land of plenty. The centrality of fertility, moreover, was not limited to cows. Indeed, it was important for women to be fertile as well—a concern not just for farmers and experts but for the settler community at large too (see chapter 5).

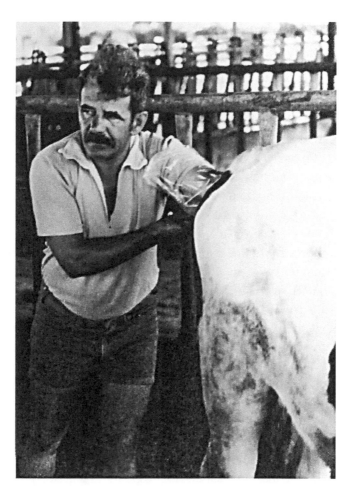

FIGURE 4.5

"Member of On [the company for cattle insemination] Inseminating a Cow at the Kibbutz," 1981. This inseminator, Arieh Shadar, later became the archivist of the collection of the Israeli Breeders Association, which is where I met him. *Source:* Kibbutz Mizra' Archives, "Cowshed Beginnings" Image Collection, Image No. 70/123.

"Our Granaries Have Filled with Grain" is a poem from the early generation of Hebrew music and lyrics written in Palestine. Based on a biblical
verse and composed in 1932–1933, the song remains famous today, and is
commonly sung during the Shavu'ot holiday to celebrate the product of
the land and milk in particular. Going beyond the minds of researchers and
breeders, the verses help illustrate how the greater population of Jewish settlers perceived the relation between the fertility of females and attempts to
create biblical plenty: "Our granaries have filled with grain and our winer-
ies with wine / Our homes are humming, humming with babies / And our
cattle is fertile / What else would you ask from us homeland / And is not
yet there? / What else would you ask from us homeland / And is not yet
there?"[112] The cries depicted are those of blessed babies, while the houses
filled with those babies stand for the fulfillment of the commitment of the
Jewish community to the land. The fertility of women, like that of cattle
and the abundance of agricultural production, all demonstrate the achievements of the Jewish settlement in Palestine. These are the components of
the ideal world for which the Jewish settlers aspired; if all went well, they
believed, the land and its creatures would be fertile. Connected here is the
fertility of women and that of cow: the birth of many cows is as important
as the birth of many babies in the attempt to satisfy the demanding land.

An interesting difference emerges, however, between the biblical text
and the ideal presented in "Our Granaries Have Filled with Grain." In Deuteronomy, plenty will be a blessing following the fulfillment of an obligation to God: "And if you faithfully obey the voice of the Lord your God. . . .
And the Lord will make you abound in prosperity, in the fruit of your womb
and in the fruit of your livestock and in the fruit of your ground, within
the land that the Lord swore to your fathers to give you."[113] As opposed to
the biblical text, the song celebrates a commitment between the people and
their land, not the people and God. Moreover, while the biblical text points
to the hand of God in shaping plenty, the song yields this responsibility to
the people. In "Our Granaries Have Filled with Grain," plenty is no longer
a godly blessing; no, in midcentury Palestine, the production of plenty has
become the obligation itself.

Since producing plenty depended on the bodies of females—both women
and cows—fertility and plenty became closely linked, and the female, the
maker of plenty, became the center of attention. Specifically, the attempts
to battle infertility in both women and cows reveal a new type of ideal

body. I am tempted to call this fertile female the "New Jewess." The bodies of those females, the New Jewesses, were shaped to become loci for the production of abundance and act as "bodyscapes" of plenty.

The search for solutions to infertility only strengthened the connection between cows and women in the efforts to create plenty in Palestine. Popular representations of this connection and crucially settlers' practices along with their search for infertility solutions demonstrate how the fertility of women became relevant. S. Freund, the first veterinary doctor employed by the Hakla'it, explained how fertility problems in cattle were handled in the 1930s. "Many theories emerged in an attempt to explain bovine infertility," he noted. "The doctor relied on the people of science from the university. Even though these were no experts in the field of veterinary medicine, they were willing to help in the study of the various problems. Among these people was Prof. Kligler from the department of bacteriology and Prof. Zondek from the department of endocrinology."[114]

Agriculturists turned to experts in human disease, such as the aforementioned microbiologist Israel Kligler (1888–1944), who is commonly remembered for his studies on malaria and public health work. While the extent of his contribution to cattle infertility is unknown, he was a frequent member of various agricultural committees from the 1920s on. In the 1930s, Kligler invested much of his energy in studying the nutrition of Jewish settlers and coinitiated the Glass of Milk a Day project, which supplied bovine milk to schoolchildren starting in 1938. It is argued that his support for the project was motivated by his will not only to better the health of children in Palestine but also to benefit dairy farming.[115] Bernhard Zondek (1891–1966) was a world-renowned sex hormone researcher and famous gynecologist in Palestine. His work focused on the problem of human infertility, and his research on gonadotropins received global appreciation (see chapter 5).

Breeders and farmers also consulted other researchers besides Kligler and Zondek. Parasitologist and physician Saul Adler (1895–1966), most famous for his work on leishmaniasis and malaria, became greatly involved in experiments on cow diseases. Veterinary doctors said that Adler brought "a new spirit," and they considered him "a scientific father and guardian."[116] This turn to researchers in the field of human medicine who were trained in medical schools in Europe became part of a larger contemporary trend. With an influx of physician immigrants and few jobs available in Palestine, many medical professionals found employment as "animal doctors."[117]

Others who trained as physicians or nurses worked with cows on farms, often simultaneously caring for sick settlers (or babies) and sick cows.[118] University researchers, together with veterinary doctors, also promoted a plan to open a joint human and animal medical school.[119] "In the scientific sense," Adler reportedly once said, "it is surprising how much veterinary and human medicine have in common. They are practically the same occupation."[120]

CENTERING REPRODUCTION

The story of the Hebrew cow allows room for the different kinds of bodies that take part in processes of place-making along with the formulation of scientific work and technological projects. Historians of Zionism have discussed how the ideal Zionist body, the New Jew, became the symbol of the nation and its manhood. But persistent investments in the fertility of the New Jewess, the maker of plenty, carried the burden of producing a sacred environment.[121] This chapter demonstrated that the efforts to create plenty applied to both humans and other animals. More specifically, we see here as well in chapter 5 how in circumstances where infertility jeopardized the production of plenty, knowledge about and practices relating to reproduction crossed species' divide.[122]

From early on, and as opposed to the European tradition, settlers' dairy cattle management focused on yield rather than looks, or as Smaragd observed admirably, "The exterior is lousy, but the yield is excellent."[123] The emphasis on the efficacy of reproductive organs correlated with the ways in which cows in Palestine/Israel were photographed—that is, from behind and as a group (in order to show the reproductive organs), rather than individually and from the side, as was common in European cattle exhibitions. By posing alone and from the side, therefore, the celebrated cow Stavit proved that she was truly exceptional (see figure 4.1).

In the story of bees, goats, and sheep (chapters 1–3), European interventions centered on changing and controlling forms of movement. Such separatism between different kinds of breeds, animals, and people was invented for and became intertwined with sustaining a new political and environmental order. In the case of the Hebrew cow, settlers manipulated bodily proximity and relied on means to overcome distance. Artificial insemination created new possibilities for production and reproduction, helping transform bovine bodies into milk-and-place-making machines. Following the 1948 war on, the composition of the Hebrew cow changed further, as wartime losses as well as new financial and technological opportunities brought thousands of US Holstein dairy cows to Israel, disrupting decades of attempts at mixing the breeds.[124]

Key political figures and scientists gathered in 1953 to lay the cornerstone for a veterinary institute in Beit-Dagan, later to become part of the Volcani Institute in honor of Elazari-Volcani. They joined in signing "the foundational scroll of building and establishment of a veterinary institute . . . that will be used as a hall of research and science for the fostering and maintenance of the health of animals in the agricultural sector in Israel." "This scroll," they wrote "is a testimony and a sign for the numerous efforts invested by the people that live in Zion in making the deserts of the land bloom, and for the revival and return of the glory of its agricultural past."[125] The bond between technoscience and the land of plenty has been institutionalized through the bodies of animals.

5 URINE AND GOLD: INFERTILITY RESEARCH AND THE LIMITS OF PLENTY

This chapter is a tale about the Hormone Research Laboratory that once existed in Jerusalem and how it became involved in creating plenty. In previous chapters, I described the attempts to create a plentiful land, tracking the literal interpretation of the "land flowing with milk and honey" metaphor, and demonstrated how those attempts shaped the physical growth of the Christian and Jewish settlement project. Along with these plans to increase particular human and animal settler populations, however, emerged a threatening problem, which exposed the limitation of the settling body: infertility. The insistence on dealing with this problem played out beyond the farm, and incorporated new sites such as the lab and clinic. It also opened the door for new participants: pregnant horses, pregnant women, and finally, aging women, whose hormone-rich urine became a key solution. In this way, the urine of mares and women was added to the honey of bees as well as the milk of cows, sheep, and white goats in the production of plenty.

This chapter explains what happened at the meeting point of global science making and local circumstances: laboratory work in the 1930s–1960s (literally) enabled the growth of the European settlement project in Palestine/Israel. As part of this same intervention, a variety of settler creatures were used, and became part of the global race for knowing hormones and their profitable production. I consider the laboratory to be a social rather than placeless institution, rooted in a particular cultural soil, and go further to show how the lab modified society.[1]

Historical studies of sex hormone research highlight the tightening relations between science and the pharmaceutical industry after World War I.
In most cases, they focus on the relationship of individual scientists with
particular pharmaceutical companies.[2] The majority of these studies analyze the development of fertility treatments for human use.[3] As this chapter
illustrates, however, the study of hormones and their production connected several locales and many individuals. More important, it was part of
a multispecies project on a most intimate scale, as knowledge about (and
the biological matter extracted from) humans became cardinal to knowledge about (and the fertility of) nonhumans, and vice versa. Furthermore,
this chapter sketches the way in which contemporary global scientific
ideas operated within a particular geopolitical context.[4] The development
of hormone research in general and emphasis on infertility in particular
were entangled with anxieties about the limits of the settlement project in
Palestine. As a result of those specific fears, fertility research developed a
local character; endocrinology "went environmental."

PRICE, MICE, AND VICE

After a long, cross-continental negotiation process, the great German gynecologist and sex hormone researcher Bernhard Zondek agreed to settle for
a monthly salary of sixty Palestine pounds, which was to be paid by the
Hebrew University (est. 1925) and US Hadassah Medical Organization. He
became an immigrant to Palestine (then under British rule).[5] It was 1934,
and Zondek had already spent many months away from his familiar Berlin environment after he was forced to leave the country as well as his
positions at the Charité Hospital and Municipality Hospital Berlin-Spandau
because he was Jewish.[6] Judah Leon (Leib) Magnes, then the chancellor of
the university in Jerusalem (and later "president"), thanked him for giving
up his hopes for higher payment, and understanding "some of the moral
and spiritual aspects of [the] situation here [in Palestine] better than some
of the other [immigrating] German scientists and doctors," who according to chancellor, were blatantly desirous of wealth. While "this might be
expected from real estate speculators, the ordinary business man and others, it was not to be expected from the scholar and the scientist," Magnes
determined, especially in times when "the country is going through . . . an
orgy of speculation and materialism."[7]

FIGURE 5.1
Professor Bernhard Zondek with three honorable women. *From the left*, his wife, Maria, their daughter, Rita, and someone who remains unknown. *Source:* Central Zionist Archives, AK/576/3.

Zondek was invited to Jerusalem in order to head the obstetrics and gynecology ward at the Hadassah Rothschild Hospital along with the parallel department at the Hebrew University's medical school that was to open that year. With his resettlement, Zondek quickly built a regional reputation, and was known as the gynecologist who treated the wealthiest and most honorable women of Jerusalem as well as the entire Middle East (see figure 5.1). But when he arrived in Jerusalem, Zondek not only intended to work at the clinic but also planned to devote much of his time to the study of hormones, and more specifically, deciphering the world of gonadotropins— sex hormones produced in the pituitary gland.[8]

He immigrated after a successful career as a practicing gynecologist, sex endocrinologist (then called sex physiologist), and inventor, and after he had already contributed immensely to the expanding hormone pharmaceutical business in Europe and the United States.[9] He is especially known for identifying the pituitary gland as the center of hormonal activity. In particular, he posited that "the anterior pituitary gland is the motor of the sexual function," and that "the gonadotropic hormone is the super-ordinated, general sex hormone under whose influence the gonadal hormone of the ovaries and testes are produced."[10] Following the discoveries of more hormones in the pituitary by other scholars, Zondek concluded that this gland "must be recognized as the conductor of the symphony of endocrine function."[11]

Zondek was also responsible for promoting urine as a substance rich in sex hormones, especially after he found traces of "female sex hormones" in the urine of pregnant women, pregnant mares, and even more so, stallions.[12] At the core of his discovery was the growing understanding that hormones can be found in and purified from biological matter, and also that hormones extracted from one creature can affect the hormonal balance and activity in another. While this general understanding of hormones created the possibility of developing fertility drugs and treatments, the reliance on bodily materials posed a great logistical and financial challenge. Zondek's discovery of sex hormones in the urine of humans and equines was revolutionary since urine was abundant, and because its availability nearly eliminated the need to search for hormones in slaughterhouses (a site from which animal ovaries, testes, and pituitary glands were obtained), gynecologic wards (for human ovaries, blood, and placentas), or morgues (for human pituitary glands). Furthermore, the possibility of extracting sex hormones from horses' urine eased the difficulties in dealing with human donors.

Zondek participated in the global standardization of sex hormones measures and was the codeveloper of the first widely used pregnancy test together with biochemist Selmar Aschheim—the Aschheim-Zondek test (or A-Z test)—in 1926.[13] This test, commonly remembered as the "rabbit test" or "frog test" (depending on the animal used in the lab), became immensely successful and was used by many thousands of women by the early 1930s.[14]

The A-Z test was based on tracking hormonal activity in the reproductive organs of animals after they were injected with the blood or urine of a pregnant female. If you inject an extract of a human woman's urine into the body of a female mouse and a particular hormonal reaction is evident in its sexual organs (in the form of bluish-black points on its ovaries), then you can conclude that the woman is pregnant. A limerick written by one of Zondek's US colleagues to honor him on his sixtieth birthday (in 1951) demonstrates the significance of this invention and the way in which the test worked, but

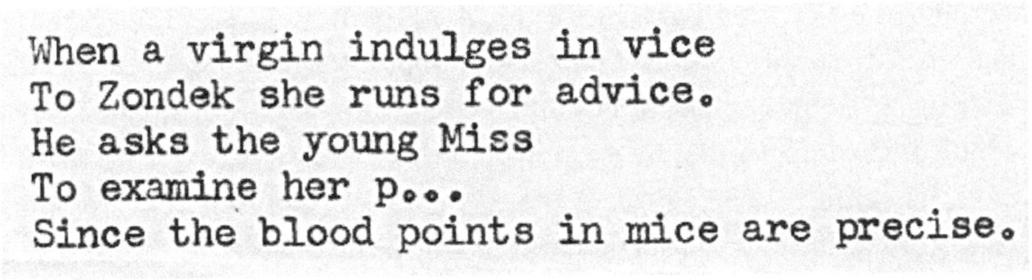

When a virgin indulges in vice
To Zondek she runs for advice.
He asks the young Miss
To examine her p...
Since the blood points in mice are precise.

FIGURE 5.2
Excerpt from a limerick composed by Bernard J. Brent of Roche-Organon pharmaceutical company in celebration of Bernhard Zondek's sixtieth birthday in 1951. *Source:* National Library of Israel Archives, Arc4°1674, 157.

also the qualities of its imagined user. The limerick reads, "When a virgin indulges in vice / To Zondek she runs for advice / He asks the young Miss / To examine her p[iss] / Since the blood points in mice are precise."[15] According to this birthday limerick, urine holds the secrets of immoral behavior, and Zondek has the power of revealing it (see figure 5.2).[16]

Zondek's research and discoveries in Europe were part of the "golden age" or "heroic age" of the reproductive sciences, or the "endocrinological gold rush," as historians defined it, peaking in the late 1920s and 1930s.[17] The making of the life sciences was undergoing a significant shift during this period. As chemical methods took center stage, and fresh tissues were favored over dead organisms, the lab (rather than the clinic) gradually gained prominence and became the most powerful center for the production of scientific knowledge.[18] In the 1920s and 1930s, scientists described the endocrinal system, discovered various biological sources from which to purify hormones, and attempted to map the reproductive system of various species. The studies conducted by Zondek and his colleagues greatly contributed to the booming market of hormonal treatments and fertility management in Europe and the United States. When he moved to Jerusalem and in spite of his expulsion from his esteemed positions, Zondek expected to continue his lab work as well as his participation in the global scientific community.[19]

TRANSLATING HORMONES IN PALESTINE

Zondek wanted to establish a laboratory for the study of hormones at the university in Jerusalem. To his great disappointment, this plan encountered significant obstacles on the way to its implementation; the university did make sure to supply a few rooms in "an Arab house" for the purpose of conducting his experiments, but not much else.[20] The house was small, Zondek claimed, so much so that he was forced to house the animals participating in his experiments, such as goats and some chickens, on the roof, creating a "farmyard atmosphere" rather than the ambience of an advanced research laboratory.[21] Worse still, both the university and hospital refused to fund the work at the lab or purchase any equipment, creating a situation where Zondek had to bear most of the financial burden himself (with the exception of a single grant from the Rockefeller Foundation).[22] It took almost six years for the university to finally agree to include the lab within its medical school. Zondek, who won numerous international prizes and honorary titles in the meantime (including repeated nominations for the Nobel Prize), finally received a budget to purchase the equipment and materials needed for daily work at the lab.[23] As such, he was able to hire a permanent assistant.[24]

Felix Gad Sulman (see chapter 3), who started working with Zondek shortly after the lab opened and was a German immigrant in his own right, at last found financial tranquility. Trained as both a "human-doctor" and "animal-doctor," Sulman was particularly suited for this lab since his multispecies qualifications facilitated his daily work with women, chickens, goats, mice, rats, and mares.[25] As contemporary laboratory work was increasingly centering on the traffic of materials across different creatures, and on the assumption that experimentation on little organisms—such as fruit flies or mice—could result in new realizations about bigger ones—such as humans—the presence of a physician veterinarian in the lab was rather useful.[26]

Sulman fit well in the lab not only because of the nature of standard biological research but also for a local reason: translation. The years in which the Zondek lab in Jerusalem operated paralleled a period of expanding European settlement in Palestine—a project that included dramatic biotic transformation. At the end of this process, new settler creatures replaced older, Indigenous ones, whether human or other. The emergence of the science of endocrinology in Palestine became deeply connected to this demographic change; as hormone therapy became a possibility, making knowledge of hormones significant to both human settlers and farm animals, Sulman's skills were crucial for translating Zondek's finding to the Palestinian context.

The act of translation was literal too, as one of Sulman's strengths was his high proficiency in Hebrew. While Zondek was required by the Hebrew University to communicate in Modern Hebrew—the designated language of settlement—this was a real struggle, which nearly prevented Zondek from participating in formal academic activities. Sulman thus wrote letters for Zondek, as indicated by his own handwriting on Zondek's letters to officials of the university, and also lectured and published in Hebrew on their lab findings.[27] Considered together, Sulman's many ways of translating Zondek—in terms of knowledge, practices, and language—were critical in mitigating global knowledge and local circumstances and demands, or to put it differently, developing a comprehensible language of knowledge in place.

Already in the first years of working at the lab, Zondek and Sulman succeeded in keeping their studies and findings in step within the aspirations of the international scientific community and its universalistic logic of objectivity. This is demonstrated by their publications (together and separately) in leading journals and academic presses that focused on biochemical investigations of hormones—that is, studies that sought to explain hormonal activity in all organisms, everywhere.[28] In the early years, the two published several articles dealing with pregnancy tests, and the purification of hormones from urine and the organs of the reproductive system.[29]

In those early years, the two were particularly occupied with a phenomenon called "antihormone," which became prominent with the emergence of hormone therapy.[30] Through this perspective, Zondek and Sulman sought to contribute to contemporary scientific conversations by explaining the gap between the growing availability of biological material for hormone purification and the treatments' low success rate.[31] Antihormone, they stressed, was a particular sensitivity to hormones, which organisms develop naturally or in reaction to hormone treatments; the injections of hormones might fail to work, according to their explanation, because bodies can develop a reaction similar to an allergy in response. By dealing with this problem, Zondek and Sulman also responded to contemporary fears regarding the move of biological matter from animals to their application in humans. In their numerous publications relating to the antihormone problem, Zondek and Sulman emphasized the reversibility of the negative effect and that endogenous hormones—that is, hormones from a human source—were always safe.[32] Moreover, Zondek contended that urine of all kinds was generally safe too, whether human or not.[33]

As the years passed, however, and while they continued to target the global hormone science community, the focus of Zondek and Sulman's studies at the laboratory gradually changed. They began dealing with other, neither organ- nor species-specific questions, but ones that were site specific; they concentrated on the problem of infertility, and more generally, examined the relation between fertility and the environment.

THE SEARCH FOR RUTIE THE HORSE: THE LAB IN THE SERVICE OF SETTLEMENT

Infertility, or sterility (terms often used interchangeably during the period), had become a severe problem in the growing European, mainly Jewish, agricultural economy already in the late 1920s. For example, in the case of dairy farming, which was positioned at the center of the settler agricultural economy, the problem was particularly troubling.[34] As discussed in chapter 4, settlers' dairy cows, which were a product of a calculated process of breeding European with local cows, suffered from infertility and a high percentage of miscarriages—a phenomenon that settler farmers considered to be a sign of "degeneration." Cow breeders noted that the Middle Eastern environment, together with the heat and efforts to produce a lot of milk, had a negative effect on the fertility of those high-quality crossbreeds. Research centers (such as the agricultural experimental stations established by the Jewish National Fund and British government) attempted to find solutions to this problem that threatened to limit milk production capacity, and the survival and growth of the settlement population as a whole.

Then hormone therapy arrived. As we have seen, the Jewish Veterinary Services consulted Zondek in its attempts to explain bovine infertility.[35] The scientists of the Hormone Research Laboratory in Jerusalem were involved in the settlers' animal farming community in various ways. During the 1940 annual meeting of the Jewish Veterinary Services, for instance, Zondek and Sulman lectured on the female reproductive system, and introduced the possibility of using hormone therapy on cattle and other farm animals. They then argued that farmers, rather than purchasing imported fertility drugs, might find it easier to get ahold of local biological material for the purpose of treating their animals. They proposed searching in slaughterhouses and hospitals, or alternatively, collecting the urine of pregnant women or mares, and then injecting it into their infertile cows in specific

FIGURE 5.3
The sheep of Kibbutz Merhavia, accompanied by a donkey. *Source:* Kibbutz Merhavia Archives Image Collection, image 142.

amounts and at specific times.[36] In addition, scientific literature from the lab's library was sent to farmers seeking solutions to infertility.[37]

Beyond catering to the needs of the professional animal farming community, Zondek and Sulman established connections and collaborations with particular farmers who were part of the expanding agricultural settlements. One of Sulman's closest collaborations was with sheep breeder and shepherd (and later painter) Aharon Harari of Kibbutz Merhavia (see chapter 3). Infertility problems that limited milk production among the sheep also threatened the existence of the herd at Merhavia as well as many others on communal agricultural settlements.[38] Harari contacted Sulman at the lab in Jerusalem in 1944 after he learned about the possibility of hormone treatments for farm animals. The two—the shepherd Harari and the doctor Sulman—quickly became colleagues and friends, and worked closely to solve fertility problems among the sheep (see figure 5.3).[39]

Documents found at the Hebrew University and Kibbutz Merhavia Archives give a sense of the material connections between the sheep-growing farm and research laboratory. Based on Zondek's early discovery of sex hormones in the urine of pregnant mares and women, and the

possibility of increasing the chances of impregnation with the injection of
sex hormones, Harari and Sulman combined forces in order to detect pregnancies in Jewish farms. The search worked as follows. First, Harari would
visit agricultural settlements where horses were raised; he would take samples of blood or urine, and send those to Jerusalem, using different means
of transportation (mainly the public bus system).[40] Sulman would then use
the A-Z test to check the blood or urine (that is, by injecting it into female
mice or rats, and examining their ovaries), and determine whether the
horse was pregnant or not.[41] Then Sulman would send a letter (or at times,
a telegraph) to Harari, noting "positive" (for pregnancy) or "negative" (for
nonpregnancy). In spring 1946, for example, Sulman wrote to Harari about
a horse named Rutie whose urine showed positive results:

> After a four-day examination, traces of prolan A were discovered in "Rutie" the
> mare.[42] Mare "Tzvia" remained negative. In the case of "Rutie" we can assume
> that she is impregnated but either at the early stages (less than 50 days) or the
> very end (after 100 days) of the pregnancy. If she is in day 40 to 50, you could
> take more serum from her and then you can be sure that you will find a lot of
> hormone in day 70.[43]

Finally, in cases where the mare (such as Rutie) was known to be pregnant, Harari would collect its urine (or blood) and inject a concentration
into his sheep in order to encourage their impregnation. When pregnant
horses were hard to find, Harari turned to another, potentially easier option
of finding pregnant women, asking for their urine, and then similarly using
it for his sheep. "For urine of women you do not need a match-maker and
you have chances of finding an easy solution. If you do not manage to get
the blood of mares, try to take a double amount of women's urine," Sulman concluded. "You can inject the serum and the urine to any clean part
of the [sheep] body, both subcutaneously and intramuscularly."[44] Sulman
provided more details about the manner in which fertility treatments in
sheep should work:

> It is recommended to inject to the sheep a first injection of 500 I.U. of prolan A.[45]
> This amount can be found in about 20cc of blood of a pregnant mare, between
> day 45 and 100 [of the pregnancy]. Since it is very difficult to determine preg-
> nancy at this stage, I do not see any other way other than for you to send the
> serum to me for examination [in the lab]. I can give you the answer in two days.
> Two weeks after the first injection you need to give them [the sheep] 500 I.U. of
> prolan B. This amount is found in 50cc of urine of pregnant women at all stages
> of pregnancy.[46]

The search for pregnant horses turned out to be more difficult than expected, as horse owners did not always cooperate, and because the horses themselves did not always conceive easily. In 1944, Sulman sent a letter to a member of the Organization of Draft, Carrier, and Riding Animals complaining that "nothing has changed since I last wrote. I understand that your stablemen are loafers of the first degree. I did not receive blood, not from [mare] Fatmeh and not from [mare] Sonia."[47] On the side of sheep farmers, however, cooperation was on the rise. As a result of his connections with the lab in Jerusalem, Harari established a reputation as a fertility expert, began instructing other farmers, and supplied hormone treatments to sheep throughout the region.[48] The shepherds of Kibbutz Kfar-Sald informed him that "impregnation has nearly seized completely, hence we ask you if you would be willing to come over and inject the hormones as promised."[49]

By so doing, Harari and many other settler farmers were able to increase the production of milk while engineering sheep pregnancies according to the time when it was in the greatest demand. Detective work at the lab and the manipulation of urine in particular played into the growing success of settlement efforts (see figure 5.4). To increase reproduction, materials as well as knowledge moved between the farm and the lab, and between the bodies of different settler creatures.

Results from such collaborations became the basis for scientific publications addressing both the local and global community. While not a qualified scientist, Harari reported to the local Jewish professional community on the different experiments he conducted using mares' and women's urine in his attempt to improve the fertility of sheep herds.[50] The names of the mares became an important part of the process: since he was experimenting with doses and combinations of donors, Harari named each concentration of urine or blood after the horse (but not the woman) from whom it originated. When he used that of a horse with the Hebrew name (of biblical origin) Rut, or Rutie as a nickname, for example, he noted using "the Rut serum." Harari reported that

> a third experiment with artificial estrus was conducted in the farm at Merhavia on 7 July 1945 on 19 sheep, 1.5 year old of age. The dose was 5cc of the "Ayala" serum. . . . [O]n 23 July 1945 they were injected a second time with different doses and serum from two different horses—"Dalia" and "Rut." 5 ewes received 10 cc of the "Dalia" serum (200 units), 3 ewes received 5cc of the "Rut" serum, and 4 ewes did not receive any because they were in heat a few days after the first injection.

FIGURE 5.4
Settler population led by a horse. (German immigrant) Eliyahu Cohen, "The Harvest Holiday March," Kibbutz Ashdot-Ya'acov, date unknown (probably late 1930s or early 1940s). *Source:* Kibbutz Ashdot-Ya'acov Archives, Eliyahu Cohen collection. Courtesy of Nadav Mann, "Bitmuna" project.

> During the 10 days following the second injection, 11 ewes were in heat, 4 of those who received 20cc of the "Dalia" serum, 3 of those who received 10cc of the "Dalia" serum, and 3 others that received the "Rut" serum.[51]

Lab results arrived after the serums were already used. They revealed the fact that Rutie was not pregnant; her serum was thus useless.

Palestine was certainly not the only place where Zondek's discoveries and techniques were used to produce fertility drugs, and in fact, Harari was not the only farmer or researcher using the bodily fluid of a horse named Rut for that purpose. In his study of the pharmaceutical company Leo and production of fertility drugs in Sweden, historian Christer Nordlund similarly observes that "Leo took its cue from the Ascheim and Zondek's discovery that female sex hormones could be extracted from the urine of pregnant mares."[52] He then details how the production of fertility drugs by Leo required massive urine collection campaigns, in which urine was sent from farms to the company lab for detection of pregnancy, and in cases of positive reply, sold to the company in great quantities. In Sweden, too, a horse named Rut donated her urine for the struggle against infertility: "The analysis of the urine from your mares gave a positive result for Rut and Alva," the company happily announced to one farmer in 1939, "from whom delivery may immediately begin."[53] As opposed to the Swedish case and other contexts studied so far, cooperation between scientists and farmers in Palestine was for a while largely independent of the pharmaceutical industry, and resulted in the direct management of fertility in the context of the farm, not the clinic. Rutie's urine traveled all the way to Jerusalem, examined through a female rodent in the lab, only to arrive at the pen and be injected into infertile sheep.

Farm needs shaped the work at the lab by way of expediting it. An article published in *Nature* in 1945 reveals how the ongoing exchange of information and materials between lab and the farm resulted in the pursuit of efficiency. In this article, titled "A Twenty-Four Hour Pregnancy Test for Equines," Zondek and Sulman proposed a technique that would give rapid results regarding horse pregnancy—that is, in twenty-four hours rather than five or six days. Without exposing the particular local need for such a test, they argued that this method is beneficial on farms since "the diagnosis of early pregnancy by a veterinarian is difficult, and . . . it is important to the breeder to recognize pregnancy in its early stage in order to prevent abortion due to excessive work or strain." They also maintained that the test could allow for the extraction of gonadotropins from mares "at the peak of hormone production."[54] In addition to such publications related to animal reproduction, Zondek published the results of his experimentation with hormone therapy in women, which he conducted as part of his gynecologic work.[55] For the scientist at the lab, the consequences of such farm-lab-clinic relations went further than generating publications, though. Gradually, settlement requirements and limitations shaped sex endocrinology, and redirected its flow.

SETTLEMENT IN THE SERVICE OF SCIENCE

Following Zondek and Sulman's own settlement in Palestine, their studies and Zondek's clinical practice gradually focused on problems of infertility, which to some degree contrasted with global trend of the post–World War II period. While scientific attention around the world turned toward family planning, the production of contraceptives, and fertility control in the Global South, as several historians of science have shown, the growing local fertility research community in Palestine/Israel concentrated on fertility promotion and finding solutions to sterility.[56] Infertility was considered a problem specific to settlers in Palestine and demanded intervention. In 1936, Zondek stated that "in Palestine, where sterility is particularly important, I have often found in sterile women a characteristic atrophy of the uterus, which is rarely seen in Europe."[57] "Sterility in women [is] becoming a serious problem," he told the *Winnipeg Tribune* in 1952, and added that a "woman's primary task is motherhood. If she doesn't fulfill it she misses part of her natural function and this can affect her physically and mentally."[58] Moreover, in an interview that took place in the year of his death, 1966, he was asked whether during his long successful career he ever dealt with the prevention of pregnancy. Zondek replied that "maybe I did some, but it was not intentional, since most of my efforts have been on behalf of fertility rather than of sterility. This was especially true after 1934 when I moved to Jerusalem and realized that sterility in certain cultural and religious contexts was a very grave problem for women in their family life."[59]

The problem of sterility was not only troubling women but as demonstrated, also tightly connected the farm to the clinic through the lab, hence offering a new meaning to Zondek's "cultural and religious contexts"—one that includes cows and sheep. Over the years, as their research projects demonstrate, Zondek and Sulman came to treat fertility differently too, in a manner that was attentive to place as opposed to organ or species. As part of their study of hormones, and in spite of the global push for standardization, these scientists attempted to understand the connection between settling creatures and a particular environment. Their studies reveal that they were increasingly interested in analyzing the effects of immigration on the body, and in a deeper sense, the consequences of Western settlement in the lands of the East.

Zondek, for example, began studying the hormonal potential of the natural environment, and examined the Dead Sea and Sea of Galilee as possible sites for the extraction of sex hormones. Using the familiar A-Z test, he injected a concentration of the water to mice and examined its influence on their reproductive organs, and was happy to discover gonadotropins in the Dead Sea—the deeper the water, the more hormones.[60] He then published the first results in *Nature* in 1937 (see figure 5.5).[61]

Œstrogenic Substances in the Dead Sea

The Dead Sea, situated at one of the deepest points of the earth, contains a high concentration of salts (more than 25 per cent). In the depth of the Dead Sea there is a sandy mud which we have found to exhibit a certain œstrogenic activity. For the mud of the southern part of the Dead Sea the œstrogenic activity is about three times as great as for the northern part. The deep sea water contains 100 m.u. per litre, while the surface water is free of œstrogenic substances. The salt manufactured from the Dead Sea (salsana) contains 100 m.u. per kgm. Our results are given in the following table.

Part of the Dead Sea	Depth	Hormone activity (m.u. per litre)
A. Water. North near Kallia	Surface	Less than 6·6
,, ,, ,,	Near the bottom	100
South	Surface	Less than 20
B. Mud	m.u. per kgm.	m.u. per kgm. dry weight
North near Kallia	30	45
South	100	100
C. Salsana		100

Other steroid hormones (male sex hormone, progesterone) could not be detected.

The mud was found to contain a yellow dye-stuff which according to its physical properties (absorption and fluorescence spectra *inter al.*) belongs to the lyochrome group.

BERNHARD ZONDEK.

Gynæcological and Obstetrical
Department, Rothschild Hadassah
Hospital, Jerusalem.

FIGURE 5.5
Bernhard Zondek's report on the discovery of sex hormones in the Dead Sea. Bernhard Zondek, "Oestrogenic Substances in the Dead Sea," *Nature* 140, no. 3536 (1937): 240.

Zondek also submitted several research proposals to the Board for Scientific and Industrial Research of the British government in Palestine to study the connection between fertility and the Palestinian environment. In one such proposal, titled the "Study of the Occurrence of Sex Hormones in Palestinian Plants," he introduced the possibility of "feeding sterile animals with plants rich in sexual hormone, thus combating sterility in domestic animals often encountered in tropical and subtropical countries," and in addition, using such plants for "the treatment of sterility in man."[62]

In another proposal, he argued that "in tropical and subtropical countries, sterility is more often encountered than in other climates. Climatic factors in reproduction have not enough been considered. . . . [T]hese studies should be carried out as well in man as in animal." As part of this study, Zondek proposed to compare cases of sterility in man and animal in Palestine and other countries with "special regard to seasonal variations," ultimately resulting in "improvement of fertility."[63] Writing in 1959 (that is, a decade after the establishment of the State of Israel), Zondek contended,

> The function of the sex glands and their relationship to fertility and sterility depends not only on endocrine function but also on environmental factors. It is widely known that climatic conditions play a considerable role; that, for instance, exposure of male animals to cold and heat results in involution of the testes and the main accessory organs with complete cessation of spermatogenesis. The female sex organs of rats react with ovarian atrophy and disturbances of the estrus cycle. In Israel, where immigration from many countries is so prevalent, we often notice amenorrhea of several months' duration in young girls unaccustomed to the subtropical climate of the Israeli coast.[64]

Sulman similarly conducted experiments on the relation between the environment and sterility. He used the ties he had established with settlers living in rural areas, such as the one with Harari at Kibbutz Merhavia, to collect data about animal fertility, but also to obtain samples of local plants and conduct experiments testing their hormonal potential. One of the plants that made him particularly curious was the mandrake, a local wild plant that had long been used to treat infertility and was commonly known as an aphrodisiac. The mandrake won great attention across folk traditions due to its magical qualities as well as because its root resembles the human body, and according to local belief, was considered to be (a small) human (see figure 5.6).[65] Sulman was interested in examining whether such fame meant that the plant contained sex hormones; he asked Harari to collect

Mandragora officinarum, L. *Tuffah el Majanin.* (Madmen's Apple).
The Mandrake. Flowers purple, leaves deep shining green. Whole
plant powerfully narcotic, but fruit has an exquisite scent, and is
eaten, though at some risk. (Half natural size).   Solanaceae).
Valley of the Convent of the Cross, Jerusalem.
Plate 73.

FIGURE 5.6
The Mandrake of Palestine, or "Madmen's Apple." *Source:* Grace M. Crowfoot and
Louise Baldensperger, *From Cedar to Hyssop: A Study in the Folklore of Plants in Palestine*
(London: Sheldon Press, 1932), plate 73. For more details on Baldensperger, her fas-
cination with Palestinian plants, and her book, see chapter 1.

samples of the plant and send them by bus to him.[66] He was finally able to
find traces of sex hormones in it. Sulman attempted to find sex hormones
in local hashish too, but found none.[67]

Also in Merhavia, Sulman and Harari conducted experiments on the
relation between light and fertility. In this study, they brought the sheep
into a relatively dark pen during different hours of the day or lit the pen
during the night, and tested how that affected their fertility.[68] Furthermore,
Sulman examined the effects of the controlled secretion of sex and other
hormones on lactation in local cows, sheep, and women.[69]

One of the main research fields that Sulman led in the post–World War
II era, and that he continued after leaving the lab, was bioclimatology (or

biometeorology)—the study of the relation between climate and bodily health. Sulman was a member of the Metrological Research Council formed by the British government, and chaired a committee that dealt with climatic effects on domestic animals. More specifically, the committee studied the relation between heat, milk production, and reproduction.[70] As was the case among cow breeders (as discussed in chapter 2 and above), the local climate bothered hormone researchers and farmers tremendously; the Middle Eastern heat was considered to have a negative effect on the health of the body and its fertility.

As the years passed, Sulman focused on the influence of heat on the human body and mental health, and dealt with cases of depression and exhaustion that, so he argued, affected people who were not used to hot weather.[71] He also devoted much attention to the question of the *sharkiye* ("wind of the East"), known as the *hamsin* too ("fifty," since it might occur up to fifty days a year), or in its Hebrew version, the *sharav*, a Middle Eastern desert wind, or heat wave phenomenon, and examined the way it affected health and hormonal balance. This relation between environment and health was further investigated in his 1976 *Health, Weather and Climate*. In the book's opening section on "the history of bioclimatology," Sulman analyzed the differences between Europeans and the people of the East:

> The people living under blue skies in sunny countries around the Mediterranean and in the Middle East, despite a lower living standard, poorer hygienic conditions and often greater prevalence of infectious diseases than in most northern countries in Europe, are usually more cheerful than the majority of the northern races who often lack the spontaneity, lightheartedness and zest of their southern brethren. On the other hand, the northern colder climates seem to give greater stimulus to the productive mental activity of man.[72]

With this adoption of essentialist assumptions about human characteristics, Sulman connected to the long tradition of colonial science, which emerged with the expansion of the European control of and settlement in non-European lands. His work presented a new face to an old scheme deemed illegitimate after World War II. Fields of study such as bioclimatology, with its deterministic aspirations—of measuring how different places create different peoples, together with what we might call "environmental endocrinology"—were nevertheless blooming among immigrant researchers in Palestine/Israel. Their bodies—always stuffed into warm, European suits—and scientific agendas were a testimony to the distress experienced in the process of settling an unfamiliar land, and moreover, the fear of the castrating East.

THE QUIET DEATH OF THE LAB

Sulman left the laboratory after fifteen years of working with Zondek, and in the late 1950s, was appointed head of the Department of Applied Pharmacology at the Hebrew University.[73] Zondek hired different researcher assistants in the following years and continued to work at the lab until his retirement in 1961, though not before selecting "a heir."[74]

In his later years, Zondek developed a new research project, which was similarly environmental in nature. He began examining the connection between sound and fertility. Realizing that "by means of the sense organs the sexual organs can be stimulated," Zondek decided to focus on the "chorionic stimulation of the auditory organ."[75] This path of research seemed promising for a while, and Zondek invested all of his energies into constructing a new acoustic space in which he would expose mice to different volumes of noise, including what he called "pure sounds," and examine how those affected their fertility.[76] Again, much like the A-Z test in which female mice were sacrificed and cut open in order to witness the influence of urine on their reproductive organs, in this study Zondek cut female mice open to see how noise affected them.[77] In order to execute his plan for a new acoustic lab, Zondek requested the support of the Hebrew University, once again, nearly thirty years later. This time, and after he was publicly celebrated as a "great doctor" and "praised researcher," the university quickly supplied him with space, speakers, and mice.[78]

This new preoccupation with noise signaled the crumbling of his long-term alliance with Sulman. Sulman, who was also a musician, choir leader, and founder of the musical festival in the Palestinian village of Abu-Gosh, and hence generally sympathetic to sound, was interfering with Zondek's attempt to create a controlled acoustic environment. In 1966, Zondek complained to the university's Territories Committee, noting that noises coming from the direction of Sulman's laboratory, now neighboring his, were ruining his experiments.[79] Soon after, without his translator, Zondek traveled to New York City to continue his investigation on the relation between acoustics and fertility after he was invited to work at the Albert Einstein Medical School.[80] He unexpectedly died there before completing the study.

The remains of the hormone research lab were dispersed across other university labs; the fate of the mice and speakers remains unknown.

CODA: "TURNING URINE INTO GOLD"

The tale of the lab shows what happened to science in the context of settlement, and the ways in which the work of immigrant scientists served and became part of a biotic settler colonial project. It was during a time when the global biomedical community learned to ignore the environment and turned to search under the skin that settler scientists decided to look sideways. Despite the logic of objectivity, universality, and placelessness positioned at the base of science making, and in spite of the attempts to participate in the global scientific community, researchers in Palestine/Israel in the mid-twentieth century were going local. By studying hormones in connection to light, heat, noise, and natural resources, their studies aligned with the properties of a new geopolitical regime. Hormone researchers gradually became interested in the relation between the problem of infertility and a particular kind of environment. In so doing, they tried to overcome an embarrassing realization: there was a great gap between their assumptions about Eastern sexuality and settler creatures' sterile reality.

Urine emerged as a savior substance. It contained the largest amount of sex hormones needed for managing infertility problems, and more important, was a resource that would never become scarce. This allowed pharmaceutical companies in Europe, the United States, and Palestine/Israel to "[find] gold in the urine."[81] Not only pharmaceutical companies profited from the manipulation of urine; so did Zondek and Sulman. Archival documents testify that both of these scientists collaborated with local pharmaceutical companies and were using the laboratory to produce hormones for commercial use.[82]

The salvaging power of urine grew even further when it was discovered that postmenopausal urine (as opposed to pregnancy urine) contains both types of sex hormones needed to induce ovulation, follicle-stimulating and luteinizing hormones, thereby enabling first pregnancies in 1961.[83] This happened at a new center for hormone research and the manipulation of urine that was established not far from Jerusalem. In the early 1960s, the endocrinology laboratory at Tel-Hashomer (today Sheba) Medical Center used the urine of postmenopausal Italian nuns and women living in elderly homes in Israel to produce—in cooperation with the pharmaceutical industry (the Italian company Serono and Israeli Ikapahrm), and the support of the Israeli prime minister and Vatican—a revolutionary drug for treating infertility named Pergonal.[84] Bruno Lunenfeld, the scientist who orchestrated this project that became a stepping-stone in the development of assisted reproductive technologies, similarly said that he was "turning urine into gold."[85] Through the lab in Jerusalem, urine connected the farm to the lab and clinic, and flowed between the bodies of different settler creatures. In the years to come, urine crossed the Mediterranean Sea as well—connecting the Israeli prime minister's office to the Vatican, elderly homes in Israel to the pharmaceutical industry in Europe, and Italian nuns to primiparous cows—gradually blurring the contours and limitations of the corporeal. An orgy of speculation and materialism indeed.

CONCLUSION: THE SYNESTHETIC EXPERIENCE

We have reached the end of the "Plenty Pageant." This book has demonstrated how modern technoscientific projects were shaped by a religious idea of the past and then became a tool for realizing that idea. Furthermore, it has described colonial state experts' and settlers' attempts to produce plenty in Palestine/Israel through the bodies of particular animals and humans, denunciation of those who did not fit their plan, and ultimate denial of the intimate forms of Indigenous knowledge on which it relied. As the different chapters have shown, these efforts centered on controlling movement—either by way of fostering or through restrictions on and managing difference. Setting the boundaries of and around producing bodies became crucial in order to deal with the threat of mixture as well as that of infertility.

The story of the production of plenty in the late nineteenth and twentieth centuries revealed various triangular structures: Bible-bees-hive, sheep-shepherd-flute, goat-peasant-land, and mares-urine-women. In this period, another fruitful bond, so to speak, formed between bees, cows, and oranges. In chapter 1, I discussed the successful relationship created between bees and citrus trees in the early twentieth century, as bees reached the Mediterranean coast, fertilized the trees, and eventually fostered a booming citrus and settler honey industry (see figure C.1). During World War II, as the movement of oranges to European markets came to a halt and the industry began to deteriorate, oranges rose to a new role: it was during this time that researchers examined the possibility of feeding oranges to cows and white goats (who were considered minicows; see chapter 2).[1]

FIGURE C.1

My grandmother, Ayala (Hinda) Novick (previously Levita), a Jewish settler from Poland in her late teens, sitting with her back to the camera on the middle right, sorting oranges before they were sent across the Mediterranean to satisfy European demand. Picture taken at a packing facility in Rehovot, Palestine, around 1940.

Prior to the crisis in orange sales, beekeeper Israel Robert Blum sent a report to the government as part of his attempts to promote beekeeping in Palestine during the British rule. That report, titled "Production of Honey in Eretz-Israel [land of Israel]," explicitly connected bees, oranges, and dairy cows, arguing that the three together could benefit the land and revive a biblical fecundity: "When we consider that the ground suitable for orange growing is still largely unplanted, that much of the greater part of the wild-yielding flowers in the hills remains unutilized, that intensive cattle raising would result in the increase of honey producing forage plants (clover) . . . we can definitely establish the fact that there is still room in Eretz Israel for hundreds of thousands of bee-hives." More time and effort were needed, he emphasized, as "it will be decades before Eretz Israel, if all its sources of honey are fully exploited, can again become, as in Bible times, a land 'flowing with honey.'"[2] Blum's report exposes his desire and expectation for great numbers. Like the shepherd David Zamir, who wanted to see numerous sheep spreading over the land (see chapter 3), and agriculturist Yitzhak Elazari-Volcani, who envisioned a million dairy cows (see chapter 4), Blum hoped for the growth of beehives to reach in the many thousands. But the decline of the citrus industry, as so often happens with such close relationships, signaled the decline of honey production.

Milk, and cow milk in particular, won central stage. The whiteness of milk played into the transmuting of power structures and its racial terms: the act of drinking milk and consuming dairy products became an important way to distinguish settlers from Indigenous people, make Jewish settlers whiter, and invent a separation between Jews and Arabs. In *Villa in the Jungle*, literary scholar Eitan Bar-Yosef examines the representation of Africa in Israeli culture, contending that Africa has been crucial to Israeli Jews because it made them seem lighter.[3] A few decades prior to the establishment of diplomatic, military, and agricultural connections between Israel and African countries, the Palestine/Israel landscape and bodyscapes were the locus of a similar "color play." As part of the settler colonial movement that enfolded this biotic transformation, the settler seemed lighter in comparison to the native. That juxtaposition of light and dark was used is both directions. Sociologist Honaida Ghanim argues that after 1948 and the Nakba, "the Palestinian person saw himself as dark, as blackish, and his color was allegoric to the color and qualities of the land; in comparison to the Jewish whiteness, the darkness (*samra*) was the proof of nativity and the unmediated bond with the land."[4]

Perceptions of animal colorings mirrored that between humans. In sharp contrast to the water buffalo and local herding black goat, the European white goat was considered rational, good-tempered, and productive. After the war, Israeli officials chose to deal with the remaining black goats by means of confinement and attempts at elimination. These creatures were a reminder of many others, both human and animal, that they believed, as environmental historian William Cronon describes about bison in colonial America, to have been "changing the color of the landscape, they were 'blackening the whole surface of the country.'"[5] Moreover, the whiteness of both the house goat and milk were important in making various types of settlers resemble each other. With the growing number of Jewish settlers

FIGURE C.2

The goat as a symbol of both Jewish and Palestinian cultural heritage, using a similar soil-like color palate. *Left*, the icon for "Yiddish," used by the Yiddish Book Center in Amherst, Massachusetts. *Right*, the icon for Al-Ma'mal ("the lab"), used by the Palestinian Foundation for Contemporary Art in Jerusalem. *Sources*: http://www .yiddishbookcenter.org/; http://www.almamalfoundation.org/index.ph.

from western and eastern Europe, the Middle East, and North Africa after 1948, white animals and white products were useful in abolishing hybrids such as Eastern-Jews and Jewish-Arabs. These whites hence aided in the construction of East-West and Arab-Jewish binaries.

But the racial separation was not complete. The white goat remained a goat, far separated from the dairy cow, the symbol of success. Designated to Jewish farmers of Middle Eastern and North African origin (the *Mizrahim*), the white goat soon began to symbolize and ridicule Mizrahi settlers themselves. In the celebrated 1964 Israeli social satire *Sallah Shabati*, for example, a white goat stands with and for the Mizrahi village fools.[6] In this way, another layer of negative meaning was added to the already-loaded animal. In past years, goats, like oranges, ultimately came to symbolize absence and loss (see figure C.2).[7]

Milk and Honey has dealt with the production of plenty, but it has said little about the use and construction of taste for these products.[8] While drinking milk and consuming soft cheeses of highly productive creatures became especially significant for the process of whitening and separation, it eventually helped create a new type of intensive nativity. This process of nativization, however, was not a singular one, as other foods contributed. Scholars have recently asked how certain foods became emblematic of Jewish Israeliness, particularly those foods with long Middle Eastern histories. Food scholar Yael Raviv, for instance, analyzes the case of falafel, and sociologist Dafna Hirsch focuses on the more recent Israeli love for hummus.[9] Like soft, white cheese, hummus has become an important component in Jewish Israeli diet and a basic product found in Israeli refrigerators. Palestinians who are citizens of Israel allegedly consume "Hebrew" dairy products too, as indicated by the derogatory name *Arab al-Shamenet* ("Cream Arabs," an amalgamation of Hebrew and Arabic) used by non-Israeli Palestinians for them (see chapter 4).[10] Like the Arab-Jewish color play, the development of such tastes moves in both directions. Side by side, these food staples potentially assemble into contested nativities—ones that are industrial, seemingly separatist, and desirably plentiful.

The tendency to utilize technologies as means for erecting plenty in Palestine/Israel goes beyond the bodies of animals and humans to incorporate plants. One such example is the long-lasting project repeatedly named, as the title of one book indicates, *The Return of the Date Palm to the Land of Israel*.[11] This project of "return" was defined in 1933 as the "efforts to introduce selected date palms in order that Palestine may regain its former fame for this fruit," and sought the cooperation between Jewish settlers, agronomists, Zionist organizations, and the British and later Israeli government.[12] The project reached its first height in the mid-1930s with the transfer (some by way of smuggling) of palm shoots from Egypt, Iraq, and Iran, and after 1948, from California as well. Jewish Agronomist Shmuel Stoller (1898–1977) was one of the key figures in these early efforts of "return." In an interview in 1976, he detailed how he used Christian Holy Land art and literature to come to understand Palestine, and attested that his scientific work of "deciphering nature" was motivated by this aspiration to "study the Bible."[13]

Like Stoller in the mid-twentieth century, Jewish American-Israeli botanist Elaine Solowey of the Arava Institute for Environmental Studies at Kibbutz Ketura currently infuses her scientific work with a combination of religious sentiments and ideas about the ancient past. Solowey has lead the movement to recover the former fame of the date fruit, using her expertise to sprout a date seed, for a while considered the "oldest known seed planted."[14] Found in the 1960s' archaeological excavations at Masada (the famous Herodian fortification destroyed in 700 CE and the site of the "Great Jewish Revolt" against Roman powers) and using radiocarbon dating, the seed was shown to be over two thousand years old, placing it "during or just before the Masada revolt."[15] In a 2008 article published in the journal *Science*, Solowey and her colleague Sarah Sallon of the Hadassah Medical Center in Jerusalem used the narrative of decline to explain the motivation for their scientific work, arguing that "the Judean Dead Sea region was particularly famous for its extensive and high-quality date culture in the 1st century CE. Over the next 2 millennia, these historic cultivars were lost, and by the early twentieth century relatively few, low quality date palms mostly propagated from seeds were recorded."[16]

"We must renew our familiarity with the ancient plants that once grew in the region and investigate them scientifically to determine their characteristics," explained Sallon in one newspaper interview.[17] Sallon gave a new interpretation to the project of return, describing how current agrotech projects aim to restore the product from the land itself rather than import it, as was the case in the 1930s: "It feels remarkable to see this seed growing, to see it coming out of the soil after 2,000 years. It's a very moving and exciting moment."[18] Ultimately, as an article from Ha'aretz noted, "the two researchers hope the reborn tree will provide valuable information about the Judean economy and society at the time of Jesus." Moreover, hinting at the parallels between palm trees and Jewish people in the project of "return to the land," the New York Times mentioned that "the date palm symbolized ancient Israel; the honey of 'the land of milk and honey' came from the date."[19]

Such technological projects operate within a particular geopolitical context. The 1930s' project of "return" was part of the larger European settler colonial project in Palestine, which is comparable to the case of the palm today. While the sprouted tree (now called Methuselah after the Bible's longest-living man) turned out to be male and hence failed to produce fruits, the project as a whole is inseparable from the larger Israeli occupation and settlement scheme within the Palestinian territories following the 1967 war. Date fruit is one of Israel's fastest-growing agricultural industries, but the areas of cultivations go far beyond the recognized borders of the state. Jewish settlers tend major areas of palm tree cultivation outside the Green Line and along the West Bank part of the Jordan Rift Valley.[20] In addition, settler agriculture in various areas of the West Bank is frequently supported by US evangelist organizations.[21]

As described, it came to be known that Methuselah was a male tree, who as recounted to me by Solowey in 2012, was "[without] a girlfriend."[22] A lonely male, the sprouted tree was unable to regain the former fame of the date fruit and revive the abundance of the ancient land (see figure C.3). Yet scientific work persisted. Solowey used "fertilizers and hormones," and recently managed to germinate six more ancient date seeds.[23] Solowey and

FIGURE C.3
Methuselah, the young palm tree that is said to have "sprouted from an ancient seed from the Masada excavations," fenced and monitored, Kubbutz Ketura, February 16, 2012. Photo taken by the author.

her colleagues consider the two females—dubbed Hannah and Judith—potential partners for Methuselah; they intend to use Methuselah in order to pollinate them.[24]

Many stories in this book expose the challenges and failures in executing the plans for plenty. Female cows, sheep, and women, for example, struggled with infertility in the 1930s and 1940s—a problem that ultimately connected the farm and clinic to the lab, and sent experts to search for solutions in bodily waste. In 1968, Israel hosted the International Congress for Fertility and Sterility, which was dedicated to the deceased hormone researcher and gynecologist Bernhard Zondek. During the closing ceremonies, Brazilian fertility expert Campus de-Paz gave a speech later portrayed as "true poetry": "People deal with fertility in this country," said de-Paz, "not only in the sense of human reproduction, but also in fertilizing the desert."[25]

Milk and Honey is arguably a local story, a tale about rural settlements in an area that is as big as New Jersey. As the various chapters illustrated, however, this process of settler colonialism was shaped by and contributed to larger regional and global processes. We have seen the ways in which beehives, cows, sheep, and white goats—just like human settlers—reached Palestine/Israel in growing numbers beginning in the late nineteenth century. But these ideas, practices, and bodies—such technologies of plenty— moved outward too. As we saw in chapter 4, the Palestine crossbred cow was known and spread across the British Empire. Furthermore, by the 1960s, planes were used to send thousands of Hebrew cows to Iranian farms. They

FIGURE C.4

An illustration attached to "Global Honey Production," a proposal by Israeli bee-keeper Israel Robert Blum (probably drawn by his brother, the artist Ludwig Blum, who used to draw the labels for his brother's jars of honey), 1976. *Source:* Central Zionist Archives, K/13/160/1.

were also used to envision (as done so by Blum) the commencement of a global system of bee movement, fertilization, and honey production (see figure C.4).[26]

No longer needed for transporting fully grown cows, planes today export their productive and reproductive potential in the form of bovine embryos. The Israeli company Maxximilk, for example, offers customers around the world the "highest quality in-vitro-ready-for-transfer-pedigree embryos" and lists four kinds of embryo products, one of which is "genetically superior, sexed, female." "It seems perfectly logical that the world's highest milk yield per cow has been achieved in Israel—'The Land flowing with Milk and Honey,'" the company details in its call to farmers, "but when you consider the heat, humidity, limited land and water resources and the plethora of veterinary issues that needed to be overcome, Israel's dairy achievements can only be described as illogical."[27] And with superior female embryos, we have come full circle. Technologies of plenty linger on.

NOTES

1. The sons of the Persian Bahá'u'lláh, the founder of the Bahá'í faith (who was sent by the Ottoman government to exile and imprisonment in Acre, Palestine), purchased Umm Djouni's lands of in the 1880s. Jewish Polish agronomist Haim Margaliot Kalvarisky then bought part of these lands on behalf of the Jewish Colonization Association in 1905.

2. Ya'akov Atzmon, "A Visit to Deganya Alef," *Cattle and Dairy Economy* 81 (1966): 2.

3. This story was recounted several times and with slight modifications in newspaper interviews with Miriam as well as articles addressing the professional cattle management community. See, for example, Irma Singer, "First Dairywoman of the First 'Kvutza,'" *Palestine Post*, February 15, 1938, 4; Zvi Lavi, "Miriam Baratz Learned How to Milk with the Bedoiuns," *Ma'ariv*, October 10, 1960, 10. The story is also included in her biography. See Smadar Sinai, *Miriam Baratz—Portrait of a Pioneer* (Ramat-Ef'al: Yad-Tabenkin, 2002), 51–52. Historian Erica Fudge discussed a similar story, taking place in early modern England, in which the combination of music, clothes, and personal familiarity were necessary for the successful milking of a cow. See Erica Fudge, *Quick Cattle and Dying Wishes: People and Their Animals in Early Modern England* (Ithaca, NY: Cornell University Press, 2018), 117.

4. Itzhak Elazari-Volcani, *The Dairy Industry as the Basis for Colonisation in Palestine* (Tel-Aviv: Palestine Economic Society, 1928). According to existing records, the village population ranged between 250 and 330 people in the 1880s. Seventy-nine Palestinian Arabs were residing in the village in 1922. A map prepared by the Palestine Jewish Colonization Association demonstrates that parts of the village lands were individually owned (by Ali Risa and Mejd el Din Jrani) and cultivated by Arabs as late as 1925. "Um el Djouni," January 7, 1926, Central Zionist Archives, J15m, 849.

5. Joseph Morgan Hodge, *Triumph of the Expert: Agrarian Doctrines of Development and the Legacies of British Colonialism* (Athens: Ohio University Press, 2007); Helen Tilley, *Africa as a Living Laboratory: Empire, Development, and the Problem of Scientific Knowledge, 1870–1950* (Chicago: University of Chicago Press, 2011), 115–168; Elizabeth R. Williams, "Cultivating Empires: Environment, Expertise, and Scientific Agriculture in Late Ottoman and French Mandate Syria" (PhD diss., Georgetown University, 2015); Jacob Norris, *Land of Progress: Palestine in the Age of Colonial Development, 1905–1948* (Oxford: Oxford University Press, 2013), 15–18.

6. Hodge, *Triumph of the Expert*, 35–36; Fredrik Meiton, *Electrical Palestine: Capitalism and Technology from Empire to Nation* (Oakland: University of California Press, 2019), 34–37.

7. Michael Bresalier, "From Healthy Cows to Healthy Humans: Integrated Approaches to World Hunger, c. 1930–1965," in *Animals and the Shaping of Modern Medicine: One Health and Its Histories*, ed. Abigail Woods, Michael Bresalier, Angela Cassidy, and Rachel Mason Dentinger (Cham, Switzerland: Palgrave Macmillan, 2018), 119–160; Jonathan Saha, "Milk to Mandalay: Dairy Consumption, Animal History, and the Political Geography of Colonial Burma," *Journal of Historical Geography* 54 (2016): 1–12; Rebecca J. H. Woods, *The Herds Shot round the World: Native Breeds and the British Empire, 1800–1900* (Chapel Hill: University of North Carolina Press, 2017). A comprehensive analysis of the importance and extent of as well as approaches to cattle and other livestock crossbreeding in different colonial contexts still awaits historians.

8. "Talk of the Fellah: There Is No Milk Left in This Cow," *Mir'at Al-Sharq*, November 22, 1930, 1.

9. On the applicability of the "settler colonialism" framework to the context of Palestine/Israel, see Areej Sabbagh-Khouri, "Tracing Settler Colonialism: A Genealogy of a Paradigm in the Sociology of Knowledge Production in Israel," *Politics & Society* 50, no. 1 (2022): 44–83; Arnon Degani, "An Invitation to Expand Zionism's Defining Contours," *Hazman Haze*, March 2019, https://hazmanhazeh.org.il/degani.

10. For the importance of the narrative of decline to French colonialism in North Africa and the British control of Iraq, see Diana K. Davis, *Resurrecting the Granary of Rome: Environmental History and French Colonialism Expansion in North Africa* (Athens: Ohio University Press, 2007); Priya Satia, "'A Rebellion of Technology': Development, Policing, and the British Arabian Imaginary," in *Environmental Imaginaries of the Middle East and North Africa*, ed. Diana K. Davis and Edmund Burke III (Athens: Ohio University Press, 2011), 23–59. On ideas about environmental decline in relation to other colonial contexts, see, for example, Richard H. Grove, *Green Imperialism: Colonial Expansion, Tropical Island Edens and the Origins of Environmentalism, 1600–1860* (Cambridge: Cambridge University Press, 1995); James Fairhead and Melissa Leach, *Misreading the African Landscape: Society and Ecology in a Forest-Savanna Mosaic* (Cambridge: Cambridge University Press, 1996).

11. Gilbert Noel Sale, "Afforestation and Soil Conservation" (lecture at the Palestine Economic Society, December 28, 1942), Israel State Archives, Mem-3, 4188, 5.

12. Sale, "Afforestation and Soil Conservation", 2; Gillian Stone, "Gilbert Noel Sale, 1897–1991," *Commonwealth Forestry Review* 70, no. 3 (1991): 88.

13. A vast literature has focused on water by way of demonstrating the relation between political power, technology, and environmental management. See, for example, Donald Worster, *Rivers of Empire: Water, Aridity, and the Growth of the American West* (New York: Oxford University Press, 1985); Wiebe E. Bijker, "Dikes and Dams, Thick with Politics," *Isis* 98, no. 1 (2007): 109–123; Chandra Mukerji, *Impossible Engineering: Technology and Territoriality on the Canal du Midi* (Princeton, NJ: Princeton University Press, 2009); Tobi Craig Jones, *Desert Kingdom: How Oil and Water Forded Modern Saudi Arabia* (Cambridge, MA: Harvard University Press, 2010); Sara B. Pritchard, *Confluence: The Nature of Technology and the Remaking of the Rhône* (Cambridge, MA: Harvard University Press, 2011); Erik Swyngedouw, *Liquid Power: Contested Hydro-Modernities in Twentieth-Century Spain* (Cambridge, MA: MIT Press, 2015).

14. On the notion of "technopolitics" and the merging of technology and political practice in the production of power structures, see Timothy Mitchell, *Rule of Experts: Egypt, Techno-Politics, Modernity* (Berkeley: University of California Press, 2002); Paul N. Edwards and Gabrielle Hecht, "History and the Technopolitics of Identity: The Case of Apartheid South Africa," *Journal of Southern African Studies* 3 (2010): 619–639. On the utility of envirotechnical analysis, see Sara B. Pritchard, "Towards an Environmental History of Technology," in *The Oxford Handbook of Environmental History*, ed. Andrew C. Isenberg (Oxford: Oxford University Press, 2014), 227–258.

15. "Palestine Honey Production," November 2, 1921, USNA, US State Department, Records of the Department of State relating to the Internal Affairs of Turkey, 1910–1929, film reel 86. I thank Samuel Dolbee for this document.

16. Dorothy Kahn, "Flowing with Honey: How It's Done in a Jewish Settlement," *Palestine Post*, August 30, 1938, 6.

17. I borrow the term "disabled histories" from Ann Stoler by way of characterizing those colonial histories that cannot be told. Ann Laura Stoler, "Colonial Aphasia: Race and Disabled Histories in France," *Public Culture* 23, no. 1 (2011): 121–156.

18. Theodore M. Porter, *Trust in Numbers: The Pursuit of Objectivity in Science and Public Life* (Princeton, NJ: Princeton University Press, 1995); James C. Scott, *Seeing Like a State: How Certain Schemes to Improve the Human Condition Have Failed* (New Haven, CT: Yale University Press, 1998).

19. Gül Şen, "The Landscape of Southern Bilād al-Shām through the Eyes of the Sixteenth-Century Ottoman Cosmographer Āşik Mehmed," in *Living with Nature and Things: Contribution to a New Social History of the Middle Islamic Periods*, ed. Bethany J.

Walker and Abdelkader Al Ghouz (Bonn: Bonn University Press, 2020), 49–78; Yaron Ben-Naeh, "'Thousands Great Saints': Evliua Çelebi in Ottoman Palestine," *Quest: Issues in Contemporary Jewish History* 6 (2013): 1–18.

20. Yaacov Shavit and Mordechai Eran, *The Hebrew Bible Reborn: From Holy Scripture to the Book of Books, a History of Biblical Culture and the Battles over the Bible in Modern Judaism* (Berlin: Walter de Gruyter, 2007), 19–20; David Norton, *A History of the Bible as Literature, Part II: From 1700 to the Present Day* (Cambridge: Cambridge University Press, 1993), 53.

21. For an example of this practice, see Philip Baldensperger, *The Immovable East: Studies of the People and Customs of Palestine* (Boston: Small, Maynard and Company, 1913), viii.

22. George Pitt, "Palestine," *British Friend* 40, no. 33 (1882): 257–258.

23. Different writers use different numbers when discussing the use of this phrase in the scriptures, from as little as sixteen to as many as thirty-one. For a discussion on the appearance and meanings of this phrase in the scriptures, see chapter 1.

24. The hymn "Jerusalem the Golden" was written by poet Bernard of Cluny in 1146 and translated to English by John M. Neale in 1858.

25. Eli'ezer Halevi, *A Journey to the Land of Israel* (1838; repr., Tel Aviv: Omanut Publishers, 1931), 21.

26. *Chronicles of the White Horse* 3 (July 1919): 11, quoted in Eitan Bar-Yosef, *The Holy Land in English Culture 1799–1917: Palestine and the Question of Orientalism* (Oxford: Oxford University Press, 2005), 280.

27. William Cronon, *Changes in the Land: Indians, Colonists, and the Ecology of New England* (New York: Hill and Wang, 1983), 6.

28. Many settlers equated the land of Palestine with ancient Israel, believed it was intimately connected to the Jewish people, and referred to it as *Eretz Yisra'el/Isruel* ("the land of Israel"). This understanding of the land is prevalent to this day.

29. Biblical descriptions influenced the development of scientific ideas and practices in many other contexts. See, for example, Janet Browne's work on the influence of the Noah's ark story on geology, and Richard Drayton's work on the influence of the biblical garden on early modern science and British imperialism. Janet Browne, *The Secular Ark: Studies in the History of Biogeography* (New Haven, CT: Yale University Press, 1983); Richard Drayton, *Nature's Government: Science, Imperial Britain and the "Improvement" of the World* (New Haven, CT: Yale University Press, 2000).

30. See, for example, Andrew D. Berns, "The Place of Paradise in Renaissance Jewish Thought," *Journal of the History of Ideas* 75, no. 3 (2014): 363; W. J. T. Mitchell, "Holy Landscape: Israel, Palestine, and the American Wilderness," *Critical Inquiry* 26, no. 2 (2002): 201, 204; Philip Mennell, *The Coming Colony: Practical Notes on Western Australia* (London: Huntington and Co., 1892), 8.

31. For a famous example demonstrating the disappointment of encountering the Palestinian reality and adoption of the narrative of decline, see Ahad Ha'am, "Truth from the Land of Israel," *Hamelitz*, June 30, 1891. The belief in the process of decline remains central to the scholarship about Palestine/Israel to this day. See, for example, Moshe Gil, "The Decline of the Agrarian Economy in Palestine under Roman Rule," *Journal of the Economic and Social History of the Orient* 49, no. 3 (2006): 285–328; Alon Tal, *Pollution in a Promised Land: An Environmental History of Israel* (Berkeley: University of California Press, 2002). Recent work has challenged this tendency, showing how ideas about the environment are rooted in political structures. Focusing on the study of water in Palestine/Israel, Samer Alatout, for instance, argues that the narrative of water scarcity emerged as late as 1948, whereas in the early twentieth century, Europeans believed that water in Palestine was abundant. Samer Alatout, "Bringing Abundance into Environmental Politics: Constructing a Zionist Network of Water Abundance, Immigration, and Colonization," *Social Studies of Science* 39. no. 1 (2009): 363–394.

32. Seminal studies of early modern science identified profound connections between Protestantism and the scientific revolution. See Max Weber, *The Protestant Ethic and the Spirit of Capitalism* (1905; repr., London: Routledge, 1992); Robert K. Merton, "Science, Technology and Society in Seventeenth Century England," *Osiris* 4 (1938): 360–632; Charles Webster, *The Great Instauration: Science, Medicine, and Reform, 1626–1660* (London: Duckworth, 1975). Some work reflects on the influence of religion in late modern society, showing that it is far from diminishing. See, for example, Talal Assad, *Genealogies of Religion: Discipline and Reasons of Power in Christianity and Islam* (Baltimore: Johns Hopkins University Press, 1993); Timothy Mitchell, ed., *Questions of Modernity* (Minneapolis: University of Minnesota Press, 2000); Charles Taylor, *A Secular Age* (Cambridge, MA: Harvard University Press, 2007).

33. For an important exception studying the intimate historical connections between Christianity and technology, see David F. Noble, *The Religion of Technology: The Divinity of Man and the Spirit of Invention* (New York: Penguin Books, 1999). Interestingly, Jennifer Karns Alexander argues that we cannot understand Jacques Ellul's critique of technological society without paying attention to its theological grounding. Jennifer Karns Alexander, "Radically Religious: Ecumenical Roots of the Critique of Technological Society," in *Jacques Ellul and the Technological Society in the 21st Century*, ed. Helena M. Jerónimo, José Luís Garcia, and Carl Mitcham (Dordrecht: Springer, 2013), 191–203.

34. See, for example, Michael Adas, *Machines as the Measure of Man: Science, Technology, and the Ideologies of Western Dominance* (Ithaca, NY: Cornell University Press, 1990); Hodge, *Triumph of the Expert*; Daniel Headrick, *Power over People: Technology, Environment, and Imperialism* (Princeton,NJ: Princeton University Press, 2009).

35. Cathy Gere, *Knossos and the Prophets of Modernism* (Chicago: University of Chicago Press, 2009); John Tresch, *The Romantic Machine: Utopian Science and Technology after Napoleon* (Chicago: University of Chicago Press, 2012); Lino Camprubí,

Engineers and the Making of the Fracoist Regime (Cambridge, MA: MIT University Press, 2014); Tiago Saraiva, *Fascist Pigs: Technoscientific Organisms and the History of Fascism* (Cambridge, MA: MIT Press, 2018).

36. See, for example, Derek Penslar, *Zionism and Technocracy: The Engineering of Jewish Settlement in Palestine, 1870–1918* (Bloomington: Indiana University Press, 1991); Tal Golan, "Introduction: Special Issue—Science, Technology, and Israeli Society," *Israel Studies* 9, no. 2 (2004): iv–viii; Meiton, *Electrical Palestine*, 45–48.

37. European settlers in Palestine thought of themselves as modern, scientific, and oftentimes secular; at the same time, they believed that land was exceptional and should be redeemed.

38. Loïc J. D. Wacquant, "Pugs at Work: Bodily Capital and Bodily Labor among Professional Boxers," *Body and Society* 1 (1995): 65–93; Debra Gimlin, "What Is 'Body Work'? A Review of the Literature," *Sociology Compass* 1, no. 1 (2007): 353–370.

39. Conevery Bolton Valencius, *The Health of the Country: How American Settlers Understood Themselves and Their Land* (New York: Basic Books, 2002); Sandra M. Sufian, *Healing the Land and the Nation: Malaria and the Zionist Project in Palestine, 1920–1947* (Chicago: University of Chicago Press, 2007).

40. Ann Laura Stoler, *Carnal Knowledge and Imperial Power: Race and the Intimate in Colonial Rule* (Berkeley: University of California Press, 2010); Geraldine Pratt and Victoria Rosner, eds., *The Global and the Intimate: Feminism in Our Time* (New York: Columbia University Press, 2012). I adopt anthropologist Hugh Raffles's "intimate knowledge" as a solution to the problems that emerge with the common use of the category of "local knowledge." See Hugh Raffles, "Intimate Knowledge," *International Social Science Journal* 173 (2000): 325–334.

41. Saraiva, *Fascist Pigs*, 14.

42. For the concept of biopower, see Michel Foucault, *The History of Sexuality, Volume I: An Introduction* (London: Allen Land, 1979).

43. The more-than-human approach, which sprouted from critical geography, is useful for writing a history that takes both humans and nonhumans into account, and focuses in particular on how human and animal lives interlaced.

44. Seminal works include Harriet Ritvo, *The Animal Estate: The English and Other Creatures in the Victorian Age* (Cambridge, MA: Harvard University Press, 1987); Donna Haraway, *Primate Visions: Gender, Race, and Nature in the World of Modern Science* (New York: Routledge, 1989); Rob Kohler, *Lords of the Fly* (Chicago: University of Chicago Press, 1994); Lorraine Daston and Gregg Mitman, *Thinking with Animals: New Perspectives on Anthropomorphism* (New York: Columbia University Press, 2006); Sarah Franklin, *Dolly Mixtures: The Making of Genealogy* (Durham, NC: Duke University Press, 2007); Emily Pawley, "Feeding Desire: Generative Environments, Meat Markets, and the Management of Sheep Intercourse in Great Britain, 1700–1750," *Osiris* 33, no. 1 (2018): 47–62.

45. Juno Salazar Parreñas, *Decolonizing Extinction: The Work of Care in Orangutan Rehabilitation* (Durham, NC: Duke University Press, 2018); Sarah Besky and Alex Blanchette, eds., *How Nature Works: Rethinking Labor on a Troubled Planet* (Albuquerque: University of New Mexico Press, 2019); Jocelyne Porcher and Jean Estebanez, eds., *Animal Labor: A New Perspective on Human-Animal Relations* (New York: Columbia University Press, 2020); Alex Blanchette, *Porkopolis: American Animality, Standardized Life, and the Factory Farm* (Durham, NC: Duke University Press, 2020).

46. See, for example, Yael Hashiloni-Dolev, *The Fertility Revolution* (Ben Shemen: Modan, 2013).

47. A desire to characterize the relation between the natural and social orders has long been a major motivation for anthropologists and historians of science. Prominent studies include Steven Shapin and Simon Schaffer, *Leviathan and the Air-Pump: Hobbes, Boyle, and the Experimental Life* (Princeton, NJ: Princeton University Press, 1985); J. Stephen Lansing, *Priests and Programmers: Technologies of Power in the Engineered Landscape of Bali* (Princeton, NJ: Princeton University Press, 1991); Drayton, *Nature's Government*. Science and technology studies scholars have attempted to complicate the nature of this relation. See, for example, Bruno Latour, *We Have Never Been Modern* (Cambridge, MA: Harvard University Press, 1993).

48. Alfred W. Crosby, *Ecological Imperialism: The Biological Expansion of Europe, 900–1900* (Cambridge: Cambridge University Press, 1986), 270; Virginia D. Anderson, *Creatures of Empire: How Domestic Animals Transformed Early America* (Oxford: Oxford University Press, 2004); Elinor G. K. Melville, *A Plague of Sheep: Environmental Consequences of the Conquest of Mexico* (New York: Cambridge University Press, 2005); Grove, *Green Imperialism*; Leo Marx, *The Machine in the Garden: Technology and the Pastoral Ideal in America* (Oxford: Oxford University Press, 1964); Diana K. Davis and Edmund Burke III, eds., *Environmental Imaginaries of the Middle East and North Africa* (Athens: Ohio University Press, 2011).

49. See, for example, Penslar, *Zionism and Technocracy*; Ilan S. Troen, *Imagining Zion: Dreams, Designs, and Realities in a Century of Jewish Settlement* (New Haven, CT: Yale University Press, 2003); Roza El-Eini, *Mandated Landscape: British Imperial Rule in Palestine, 1929–1948* (New York: Routledge, 2006); Hezi Amiur, *Mixed Farm and Smallholding in Zionist Thought* (Jerusalem: Zalman Shazar Center, 2016). One exception is sociologist Saul Katz's work, which pays attention to agricultural and technological practices. See, for example, Saul Katz, "'The First Furrow': Ideology, Settlement, and Agriculture in Petah-Tikva in Its First Decade," *Catedra* 23 (1982): 124–157; Saul Katz, "Sociological Aspects in the Development (and Replacement) of Agricultural Knowledge in Israel: The Emergence of Ex-Scientific Systems for the Production of Agricultural Knowledge, 1880–1940" (PhD diss., Hebrew University, 1986).

50. See, for example, Huri İslamoğlu, "Property as a Contested Domain: A Reevaluation of the Ottoman Land Code of 1858," in *New Perspectives on Property and Land in the Middle East*, ed. Roger Owen (Cambridge, MA: Harvard University Press, 2000):

3–61; Gershon Shafir, *Land, Labor and the Origins of the Isræli-Palestinian Conflict, 1882–1914* (Cambridge: Cambridge University Press, 1989), 32–34; Ya'akov Firestone, "Crop-Sharing Economics in Mandatory Palestine—Part I," *Middle Eastern Studies* 11, no. 1 (1974): 3–23; Amos Nadan, "Colonial Misunderstanding of an Efficient Peasant Institution: Land Settlement and Mushā' Tenure in Mandate Palestine, 1921–47," *Journal of Economic and Social History of the Orient* 46, no. 3 (2003): 320–354.

51. Edward Said, *Orientalism* (New York: Pantheon Books, 1978); Lester I. Vogel, *To See Promised Land: Americans and the Holy Land in the Nineteenth Century* (University Park: Penn State University Press, 1993); Haim Goren, *Go Research the Land: German Studies of the Land of Israel in the Nineteenth Century* (Jerusalem: Yad Ben-Zvi, 1999); John Jams Moscrop, *Measuring Jerusalem: The Palestine Exploration Fund and British Interest in the Holy Land* (London: Leicester University Press, 2000); Gil Eyal, *The Disenchantment of the Orient: Expertise in Arab Affairs and the Israeli State* (Stanford, CA: Stanford University Press, 2006).

52. Hillel Cohen, *1929: Year Zero of the Jewish-Arab Conflict* (Jerusalem: Keter, 2013); Baruch Kimmerling and Joel S. Migdal, *Palestinians: The Making of People* (Cambridge, MA: Harvard University Press, 1994), 102–134; Mahmoud Yazbak, "From Poverty to Revolt: Eeconomic Factors in the Outbreak of the 1936 Rebellion in Palestine," *Middle Eastern Studies* 36, no. 3 (2000): 93–113; Ted Swedenburg, *Memories of Revolt: The 1936–1939 Rebellion and the Palestinian National Past* (Fayetteville: University of Arkansas Press, 2003).

53. Benny Morris, *The Birth of the Palestinian Refugee Problem, 1947–1949* (Cambridge: Cambridge University Press, 1987); Shira Robinson, *Citizen Strangers: Palestinians and the Birth of Israel's Liberal Settler State* (Stanford, CA: Stanford Univesity Press, 2013).

54. Stoler calls for reading the archives along their grain, before reading against them, as colonial studies scholars have tended to do. Reading along the colonial archives' grain, she argues, is essential for analyzing and comparing colonial rules as well as their texture, circuits of knowledge production, and racial commensurabilities. Ann Laura Stoler, *Along the Archival Grain: Epistemic Anxieties and Colonial Common Sense* (Princeton, NJ: Princeton University Press, 2010), 17–54.

INTERLUDE

1. Amnon Cohen and Bernard Lewis, *Population and Revenue in the Towns of Palestine in the Sixteenth Century* (Princeton, NJ: Princeton University Press, 1978), 3–12, 42–45.

2. This changed in 1885, when taxes were also imposed on camels and cows, and changed further in 1903 with taxes on horses, mules, and donkeys as well. See Farid

Al-Salim, *Palestine and the Decline of the Ottoman Empire: Modernization and the Path to Palestinian Statehood* (London: Bloomsbury, 2015), 145. The Ottoman Animal Law of 1905 made annual stock counts compulsory. See Roza El-Eini, *Mandated Landscape: British Imperial Rule in Palestine, 1929–1948* (New York: Routledge, 2006), 226. Historical geographers Wolf-Dieter Hütteroth and Kamal Abdulfattah argue that the explanation for the joint and consistent tax on goats and beehives is that "both goats and bees must have been looked upon as serving the most essential needs— namely as supplying milk and sugar—and they were therefore taxed consistently all over the country." Wolf-Dieter Hütteroth and Kamal Abdulfattah, *Historical Geography of Palestine, Transjordan and Southern Syria in the Late 16th Century* (Erlangen, Germany: Franconian Geographic Society, 1977), 72. Animals were also taxed as food as part of the urban slaughterhouse activity. On the growth of this activity in Palestine in the eighteenth century, see Amnon Cohen, *Economic Life in Ottoman Jerusalem* (Cambridge: Cambridge University Press, 1989).

3. Hütteroth and Abdulfattah, *Historical Geography of Palestine*, 3, 6.

4. On this change from knowledge by kind to monetary value as the basis of tax evaluation, see Anthony Greenwood, "Istanbul's Meat Provisioning: A Story of the Celepkeşan System" (PhD diss., University of Chicago, 1988), 248. For a discussion about the difference between knowledge of animals by kind and knowledge of production value, see Huri İslamoğlu-İnan, *State and Peasant in the Ottoman Empire: Agrarian Power Relation and Regional Economic Development in Ottoman Anatolia during the Sixteenth Century* (Leiden: Brill, 1994), 41.

5. Historians of the Ottoman Empire have debated the nature of these changes for a long time, and in the last three decades, attempted to challenge the earlier and widespread declensionist narratives. Rather than characterizing these narratives as a decline of the empire, such scholars explained this period as one of transformation, privatization, or (environmental) crisis. See, respectively, Baki Tezcan, *The Second Ottoman Empire: Political and Social Transformation in the Early Modern World* (Cambridge: Cambridge University Press, 2010); Ariel Salzmann, "An Ancien Régime Revisited: 'Privatization' and Political Economy in the Eighteenth-Century Ottoman Empire," *Politics and Society* 21, no. 4 (1993): 393–423; Sam White, *The Climate of Rebellion in the Early Modern Ottoman Empire* (New York: Cambridge University Press, 2011).

6. Beshara Doumani, *Rediscovering Palestine: Merchants and Peasants in Jabal Nablus, 1700–1900* (Berkeley: University of California Press, 1995), 41–44, 25–26.

7. Amnon Cohen, "Ottoman Rule and the Re-Emergence of the Coast of Palestine (17th–18th Centuries)," *Revue de l'Occident Musulman et de la Méditerranée* 39 (1985): 163–175; Beshara B. Doumani, "Review of Thomas Philipp, Acre: The Rise and Fall of a Palestinian City, 1730–1831," *Journal of Palestine Studies* 33, no. 1 (2003): 98–100; Doumani, *Rediscovering Palestine*, 31–32, 105; Alexander Schölch, "The Economic Development of Palestine, 1856–1882," *Journal of Palestine Studies* 10, no. 3 (1981):

35–58; Haim Gerber, "Modernization in Nineteenth-Century Palestine: The Role of Foreign Trade," *Middle Eastern Studies* 18, no. 3 (1982): 250–264; Nahum Karlinsky and Mustafa Kabha, "The Missing Orchard: Palestinian-Arab Citrus Cultivation before 1948," *Zmanim: A Historical Quarterly* 129 (2015): 94–109.

8. For a discussion on the importance of goats, sheep, and cows to village life on the edge of northeastern Palestine during the Ottoman period, see Martha Mundy and Richard Saumarez Smith, *Governing Property, Making the Modern State: Law, Administration, and Production in Ottoman Syria* (London: I. B. Tauris, 2007), 208–232.

9. Amy Singer, *Palestinian Peasants and Ottoman Officials: Rural Administration around Sixteenth-Century Jerusalem* (Cambridge: Cambridge University Press, 1994), 51. On the Ottoman "animal tax" (*aǧnam rüsumu*, literally "sheep tax"), which usually included sheep and goats, see Nora Barakat, "Marginal Actors? The Role of Bedouin in the Ottoman Administration of Animals as Property in the District of Salt, 1870–1912," *Journal of the Economic and Social History of the Orient* 58, no. 1–2 (2015): 105–134.

10. For an example of the European interest in documenting these practices at the turn of the twentieth century, see Philip J. Baldensperger, "Women in the East," *Palestine Exploration Fund Quarterly Statement* (1900): 66–67.

11. My calculation is based on survey records that appear in Hütteroth and Abdulfattah, *Historical Geography of Palestine*. The lowest possible count is 63,905, and highest is 66,205 (in two villages, the count also included goats, making it impossible to know how much was paid for the individual species). Akçe is a small silver coin, the chief Ottoman monetary unit of the Ottoman Empire.

12. Singer, *Palestinian Peasants and Ottoman Officials*, 49; Hütteroth and Abdulfattah, *Historical Geography of Palestine*, 43. Hütteroth and Abdulfattah's precise estimation is 206,290 people. Geographer Shukri Arraf also used these records to estimate the number of water buffalo, and his calculus is not that different: 13,906. Shukri Arraf, *The Sources of the Palestinian Economy from the Earliest Periods and until 1948* (Ma'aliya, Sudan: Dar El Amaq, 1997), 190–194.

13. Zohar Amar and Yaron Serri, "When Did the Water Buffalos Arrive to the Water Landscapes of the Land of Israel?," *Katedra* 117 (2005): 63–70. Burchard Brentjes uses archaeological evidence to argue for an earlier arrival of water buffalo to the area. Burchard Brentjes, "Water Buffalo in the Cultures of the Ancient Near-East," *Zeirschrift für Saugetierkunde* 34 (1969): 187–191.

14. Shihāb al-Dīn al-Nuwayrī, *The Ultimate Ambition in the Arts of Erudition: A Compendium of Knowledge from the Classical Islamic World* (New York: Penguin Books, 2016), 163. Also quoted in Amar and Serri, "When Did the Water Buffalos Arrive?," 70.

15. Alan Mikhail, *The Animal in Ottoman Egypt* (Oxford: Oxford University Press, 2013), 22–24.

16. Yehuda Karmon, "The Drainage of the Huleh Swamps," *Geographical Review* 50, no. 2 (1960): 173.

17. Teodore Larsson, "A Visit to the Mat Makers of Huleh," *Palestine Exploration Fund Quarterly Statement* 68, no. 4 (1936): 225–230.

18. Sliman Khawalde and Dan Rabinowitz, "Race from the Bottom of the Tribe That Never Was: Segmentary Narratives amongst the Ghawarna of Galilee," *Journal of Anthropological Research* 58, no. 2 (2002): 225–243. The celebrated 1998 novel *Gate of the Sun* (*Bab al-Shams*), written by Palestinian author Elias Khoury, includes a Palestinian grandmother who tells her grandson a story about a water buffalo breeder, noting that "the Bedouins raise water buffalos, not us [peasants]." Elias Khoury, *Bab al-Shams* (Beirut: Dar el Adab, 1998), 361.

19. This fable was documented with slight variations by both Stephan Hanna Stephan and A. M. Spoer. See Stephan Hanna Stephan, "Palestinian Animal Stories and Fables," *Journal of the Palestine Oriental Society* 3 (1923): 185–186; A. M. Spoer, "Palestine Folktales," *Folklore* 42, no. 2 (1931): 155–156. Describing this fable, Spoer concludes by noting that "I considered this story the most immoral I had heard, and quite characteristic of the money-grabbing fellahin [peasants]."

20. Rakan Mahmoud, "Oral History Interview with Warda al-Abdallah," Palestine Remembered, March 10, 2010, https://www.palestineremembered.com/Safad/Mallaha /Story17992.html (minutes 2:00:00–2:04:00).

21. Muntaha Lubani and Ibtisam Abu Salim, "Interview with Muhammad Qasim Muhammad," Palestinian Oral History Archive, American University of Beirut Libraries, October 30, 1998,https://libraries.aub.edu.lb/poha/Record/4695 (minutes 0:27:30–0:29:24).

22. Rakan Mahmoud, "Oral History Interview with Ahmad Ismail Dakhloul," Palestine Remembered, January 18, 2010, https://www.palestineremembered.com/Safad /al-Salihiyya/Story17996.html (minutes 2:18:00–2:36:00).

23. The 1922 *Handbook of Palestine* noted that 615 living and 63 slaughtered buffalo were counted in the 1920–1921 census. Harry Charles Luke and Edward Keith Roach, eds., *The Handbook of Palestine, Issued under the Authority of the Government of Palestine* (London: Macmillan and Co., 1922). The British director of customs, excise, and trade remarked in 1927 that "according to revenue returns, there are in Palestine . . . 4,457 buffaloes." K. W. Stead, *Report on the Economic and Financial Situation of Palestine* (London: H. M. Stationery Office, 1927), 20.

24. Captain I. Gillespie, "Livestock Survey—Palestine & Transjordan," August 30, 1943, National Archives of the UK, FO 922, 72, 2. Gillespie reports that "a livestock census was carried out in Palestine in 1937 and the following is a comparison with the present enumeration in round numbers: buffaloes: 1942—4,000, 1937—6,000." I thank Efrat Gilad for this document.

25. Comparing early Ottoman to British records, which indicate that buffalo can only be found near the Huleh, it seems that their habitat shrunk substantially. In addition, the 1596–1597 Ottoman survey includes a count of three villages with the name Jamaseen (literally meaning "the people water buffalo") —indicating that their inhabitants raised and bred buffalo. British records indicate the existence of two neighboring villages, eastern and western Jamaseen, in the surroundings of the Auja River (currently the Hayarkon River); they were both ruined during the 1948 war.

26. William McClure Thomson, *The Land and the Book; or, Biblical Illustrations Drawn from the Manners and Customs, the Scenes and Scenery, of the Holy Land* (London: T. Nelson and Sons, 1859), 389, 251.

27. In 1934, the British government transferred the concession over Huleh from the Syro-Ottoman Agricultural Company into Zionist hands (the Palestine Land Development Company). Hence British drainage plans expedited and intensified older interventions in the area.

28. Rendel, Palmer and Tritton, *Huleh Basin: Report, Preliminary Scheme and Estimate for Reclamation of the Huleh Lake and Marshes and Drainage and Irrigation of the Basin* (London: Palestine EMICA Association, July 1936), Israel State Archives [hereafter ISA], Mem-1, 4431, 30. Also quoted in Sandra M. Sufian, *Healing the Land and the Nation: Malaria and the Zionist Project in Palestine, 1920–1947* (Chicago: University of Chicago Press, 2008), 301.

29. Solicitor general, Attorney General's Office, Jerusalem, to district commissioner, Galilee District, April 2, 1943, ISA, Mem-29, 471, 1. On the history of goat grazing restrictions, see chapter 2.

30. Gillespie, "Livestock Survey," 5.

31. Sufian, *Healing the Land and the Nation*, 1.

32. The Israeli Hula Nature Reserve was established in 1964. Four decades after the completion of Huleh's drainage, an artificial lake (Agmon Hahula) was created and became a globally renowned site for bird migration watching. A program to reintroduce water buffalo began in the early 2000s—a time in which seven local female buffalo allegedly survived. A first crossbred calf of local and Italian parents was born there in 2007. See Eli Ashkenazy, "The Water Buffalos Are Coming Back: A New Buffalo Born in the Huleh Valley," *Ha'aretz*, June 25, 2007, https://www.haaretz.co.il/misc/article-print-page/1.1420274.

CHAPTER 1

1. For the importance of the Holy Land to British culture and during the British presence in Ottoman Palestine, see Eitan Bar-Yosef, *The Holy Land in English Culture 1799–1917: Palestine and the Question of Orientalism* (Oxford: Oxford University Press, 2005). For US Holy Land tourism, see Lester I. Vogel, *To See Promised Land:*

Americans and the Holy Land in the Nineteenth Century (University Park: Penn State University Press, 1993).

2. Frank Benton, "The New Races of Bees," *American Bee Journal* 20, no. 3 (1884): 38; Eva Crane, *The World History of Beekeeping and Honey Hunting* (New York: Routledge, Chapman, and Hall, 1999), 369.

3. James P. Strange, "A Severe Stinging and Much Fatigue—Frank Benton and His 1881 Search for Apis Dorsata," *American Entomologist* 47, no. 2 (2001): 112–116; Andrew H. Divan, "First Queen by Mail from Jerusalem," *American Bee Journal* 20, no. 51 (1884): 809. There are some indications showing that the US hive was not the only movable frame beehive to arrive in Palestine in 1880. Haim Goren and Richab Rubin argue that the German Templar settlers used the European Dzierzon hive in settlements in the Haifa area. Haim Goren and Richab Rubin, "This Is How Modern Beekeeping Started in This Land," *Mada* 29, no. 4–5 (1985), https://web.archive.org /web/20140907225240/http://www1.snunit.k12.il//heb_journals/mada/294181.html.

4. The St. Chrischona Pilgrim Mission was established at 1840.

5. Alex Carmel, "C. F. Spittler and the Activities of the Pilgrims Mission in Jerusalem," in *Ottoman Palestine, 1800–1914: Studies in Economic and Social History*, ed. Gad G. Gilber (Leiden: E. J. Brill, 1990), 255, 256, 270.

6. Carmel, "C. F. Spittler," 272.

7. Philip Baldensperger, *The Immovable East: Studies of the People and Customs of Palestine* (Boston: Small, Maynard and Company, 1913), viii, x. Partially due to its proximity to Bethlehem and ancient pools known as Solomon's Pools, Artas became a center of European exploration and settlement throughout the nineteenth century. Among the Europeans who lived in Artas were James Finn, the British consul of Jerusalem (1846–1863), who bought a house in the village; British missionary John Meshullam (1799–1878), who attempted to establish an agricultural settlement there; and anthropologist Hilma Granqvist (1890–1972), who lived with the Baldenspergers.

8. In 1942, Philip published this information about his father's training and noted that Henri was deeply saddened by Jadallah's death in 1869. Falestine Naili, "Henri Baldensperger: "Henri Baldensperger: An Alsatian Missionary and 'Living Together' in Ottoman Palestine," *L'Annuaire de la Société d'Histoire de la Hardt et du Ried* 23 (2011): 14.

9. Crane, *The World History of Beekeeping and Honey Hunting*, 175–176, 269.

10. Jean Baldensperger, "Palestine: An Account of Bee-Keeping There by an Eye-Witness," *American Bee Journal* 24, no. 4 (1888): 60.

11. A'lia later became Hilma Granqvist's main informant in her work on the people of Artas. See Hilma Granqvist, *Marriage Conditions in a Palestinian Village* (Helsinki: Centraltryckeri och Bokbinderi, 1931); Hilma Granqvist, *Birth and Childhood among the Arabs: Studies in a Huhnmadan Village in Palestine* (Helsinki: Söderström ja Co Förlagsaktiebolag, 1947).

12. Ada Goodrich-Fereer, *Arabs in Tent and Town: An Intimate Account of the Family Life of the Arabs of Syria, Their Manner of Living in Desert and Town, Their Hospitality, Customs and Mental Attitude, with a Description of the Animals Birds, Flowers and Plants of Their Country* (New York: G. P. Putnam's Sons, 1924), 73.

13. Quoted in Goodrich-Fereer, *Arabs in Tent & Town*, 60.

14. Baldensperger, *The Immovable East*, x.

15. Carmel, "C. F. Spittler," 272.

16. Philip's articles, titled "The Immovable East," later became a permanent column in the *Palestine Exploration Fund Quarterly Statement*. The collection of these articles was published in 1913 as a book under the same name. John Jams Moscrop, *Measuring Jerusalem: The Palestine Exploration Fund and British Interest in the Holy Land* (London: Leicester University Press, 2000); Nadia Abu El-Haj, *Facts on the Ground: Archaeological Practice and Territorial Self-Fashioning in Israeli Society* (Chicago: University of Chicago Press, 2001).

17. Philip Baldensperger, "The Identification of Ain-Rimmon with Ain-Urtas (Artas)," *Palestine Exploration Fund Quarterly Statement* 43 (1912): 201–2011.

18. Baldensperger, "Palestine," 59–60.

19. Granqvist, *Marriage Conditions in a Palestinian Village*), 9.

20. Timothy Mitchell, *Rule of Experts: Egypt, Techno-Politics, Modernity* (Berkeley: University of California Press, 2002); Gabrielle Hecht, "Rupture-Talk in the Nuclear Age: Conjuring Colonial Power in Africa," *Social Studies of Science* 32, no. 5–6 (2002): 691–727.

21. See, for example, Michael Adas, *Machines as the Measure of Man: Science, Technology, and the Ideologies of Western Dominance* (Ithaca, NY: Cornell University Press, 1990), 7, 22.

22. Jean Comaroff and John L. Comaroff, *Of Revelation and Revolution, Volume 1: Christianity, Colonialism, and Consciousness in South Africa* (Chicago: University of Chicago Press, 1991), 5–6. On the role of missionaries as agents of geopolitical transformations, see Heather J. Sharkey, *American Evangelicals in Egypt: Missionary Encounters in an Age of Empire* (Princeton, NJ: Princeton University Press, 2008). For missionaries in Palestine, see, for example, Ruth Kark, "Millenarism and Agricultural Settlement in the Holy Land in the Nineteenth Century," in *Deutsche in Palästina und ihr Anteil an der Modernisierung des Landes*, ed. Jacob Eisler (Wiesbaden: Harrassowitz Verlag, 2008), 14–29; Carmel, "C. F. Spittler," 265.

23. Henry S. Osborn, *Palestine: Past and Present, with Biblical, Literary and Scientific Notices* (Philadelphia: James Challen and Son, 1859), 506–507.

24. Baldensperger, "Palestine," 60. As demonstrated below, Jean's reference to the "holy bee" was not unique. The global apiary community used this term prior to Jean's publication. This undermines the claim that the Baldenspergers were an atypical case.

25. Grace M. Crowfoot and Louise Baldensperger, *From Cedar to Hyssop: A Study in the Folklore of Plants in Palestine* (London: Sheldon Press, 1932), 59. The book was written by Crowfoot, and was based on Louise's knowledge and experience. Crowfoot (1877–1957) was primarily a British archaeologist and explorer of Egypt, but also had worked in Sudan and Palestine as part of the Palestine Exploration Fund. She was particularly interested in textiles, weaving, and hand-spinning techniques.

26. Strange, "A Severe Stinging and Much Fatigue," 112–113.

27. Benton, "The New Races of Bees," 38–39. Benton and Jones were not the only ones advocating for the value of the holy bee in the United States. See, for example, H. Alley, "The Holy Land Bees," *American Bee Journal* 20, no. 37 (1884): 586. In his famous beekeeping manual, US apiculturist E. F. Phillips (1878–1951) mentioned this journey and its results. He counted two kinds of bees in Palestine: one identical to the Egyptian species, and one that was "introduced into America in 1880 by Jones and Benton but were soon abandoned as valueless." E. F. Phillips, *Beekeeping*, ed. Liberty Hyde Baily (New York: Macmillan Company, 1914), 195.

28. Several movable frame beehive models were invented and tested concurrently in Europe and North America during the nineteenth century. The Langstroth movable frame beehive, which ultimately won the greatest success around the world, was invented and patented by US reverend L. L. Langstroth in 1852. See Crane, *The World History of Beekeeping and Honey Hunting*, 422.

29. Straw basket hives, which were common in Europe, were not fixed, but they did not have frames.

30. For an analysis of technologies as systems, see Thomas P. Hughes, *Networks of Power: Electrification in Western Society, 1880–1930* (Baltimore: Johns Hopkins University Press, 1993). On the interactions between various kinds of workers, both Jews and Arabs, in the construction of railways in British Palestine, see Zachary Lockman, *Comrades and Enemies: Arab and Jewish Workers in Palestine, 1906–1948* (Berkeley: University of California Press, 1996), 812. Lockman focuses on railway workers in order to challenge the "dual society" paradigm in the historiography of British Palestine, which separates the Jewish and Arab communities.

31. Crowfoot and Baldensperger, *From Cedar to Hyssop*, 59.

32. Beyond Lockman's book, little work focuses on railroads in Palestine throughout the nineteenth and first half of the twentieth centuries. Yair Safran and Tamir Goren, for example, write about plans to construct railroads in late Ottoman northern Palestine. Yair Safran and Tamir Goren, "Ideas and Plans to Construct a Railroad in Northern Palestine in the Later Ottoman Period," *Middle Eastern Studies* 46, no. 5 (2010): 753–770. Furthermore, almost no work revolves around other transport technologies. Several popular accounts mention that cars were extremely rare in Palestine prior to World War I, yet became more prevalent as the war progressed. For instance, according to the written media, cars were owned by a rich few in the 1920s, but became increasingly common in the 1930s. In 1933, there were only six

thousand cars in Palestine, but approximately ten thousand by 1935. "A Car for Every One Hundred Thousand People in the Land of Israel," *Doar Hayom*, February 25, 1935, 1. Lockman (*Comrades and Enemies*, 186–187) argues that "motor transport had developed very quickly in Palestine during the late 1920s and the early 1930s as the government built new roads and improved existing ones." On road-paving schemes and their relation to partition plans under the British rule, see Shira Pinhas, "Road, Map: Partition in Palestine from the Local to the Transnational," *Journal of Levantine Studies* 10, no. 1 (2020): 111–121. On the emergence of the Jewish flight culture in British Palestine, see Yossi Malchi, "Modernity, Nationality, and Society: The Beginning of Hebrew Flight in Palestine 1932–1940" (MA thesis, Tel Aviv University, 2007).

33. Edgerton examines the old that is part of the utilization of new technologies as well as the change in technologies over time. In so doing, he argues against the centrality of inventions and inventors in historical analyses along with the equation of technology with progress. Edgerton also coins the term "creole technologies" to refer to the technologies of the poor and the creative utilization of whatever parts are available for making technologies work—technologies of "making do." Using the broader meaning of "creole," I adopt the term to analyze technologies of mixture. David Edgerton, *The Shock of the Old: Technology and Global History since 1900* (Oxford: Oxford University Press, 2007).

34. Saleh Merrill, "Honey Producing in Old Palestine," *American Bee Journal* 40, no. 23 (1900): 356. For the temporary movement of beehives on wagons drawn by horses, see Clifford M. Zierer, "Migratory Beekeepers of Southern California," *Geographical Review* 22, no. 2 (1932): 260–261.

35. For a discussion of the problem of camels and other animals in disturbing the movement and functionality of trains and railroads, see Lydda district engineer sgd. F. H. Taylor to chair of the Soil Conservation Board of the British Mandate in Palestine, July 11, 1942, Israel State Archives [hereafter ISA], M-13, 5109.

36. Crowfoot and Baldensperger, *From Cedar to Hyssop*, 59.

37. Crowfoot and Baldensperger, *From Cedar to Hyssop*, 59–60. Similar descriptions are found in the records of the Lerrer family from Ness Ziona. This Jewish family adopted the movable frame beehive directly from the Baldenspergers in the late 1880s. Israeli Honey Council, "On the History of Beekeeping in Israel: Problems and Related Stories," http://www.honey.org.il/info/about-us/begin_peer.htm.

38. Frank Benton, "A Bee-Convention in Syria," *American Bee Journal* 21, no. 35 (1885): 551.

39. Baldensperger, *The Immovable East*, xi. In spite of Philip's claim, the eldest of the Baldensperger family, Theophile, did not take part in the family beekeeping practice, as he left Palestine as early as 1865 and settled in the French Saint-Dié-des-Vosges. Captain Egon Potter, "Ph. J. Baldensperger's Career," *Bee World* 8 (1927): 156–157. According to records of the Ecole Nationale d'Ingeniéur, Sud Alsace, Mulhouse,

Theophile finished his education there in 1874 and therefore was probably an engineer rather than a beekeeper. Ecole Nationale d'Ingeniéur, Sud Alsace, Mulhouse, "Graduates," http://www.anciens-ensisa.org.

40. Granqvist, *Birth and Childhood among the Arabs*, 20.

41. Baldensperger, *The Immovable East*, xi. I have found no record of Jean's life circumstances beyond Philip's accounts, thereby leading me to conclude that Jean died soon after his return to Palestine.

42. Baldensperger, *The Immovable East*, xii; Philip Baldensperger, "Marketing Honey on the Shores of the Mediterranean," *Bee World* 11 (1930): 133–136.

43. As part of his work, Philip established the Société d'Apiculture des Alpes Maritime in 1921 and its bulletin. He participated and represented France in Apiculture International congresses, and became the president of the Apis Club in 1927. For further details about Philip's career, see E. F. Phillips, "Mr. P. J. Baldensperger," *Bee World* 8 (1927): 97–99; Egon Potter, "Ph. J. Baldensperger's Career," *Bee World* 8 (1927): 156–157; Egon Potter, "Ph. J. Baldensperger," *Bee World* 29 (1948): 73.

44. "Obituary Notices: Nora Baldensperger," *Bee Journal* 58 (1977): 128.

45. Baldensperger, *The Immovable East*, xii.

46. Granqvist, *Birth and Childhood among the Arabs*, 171.

47. Granqvist, *Birth and Childhood among the Arabs*. On the history of the citrus industry in Palestine, relation between Palestinian and Jewish techniques, and gradual Jewish domination of the industry, see Nahum Karlinsky, *California Dreaming: Ideology, Society, and Technology in the Citrus Industry of Palestine 1890–1939* (Albany: SUNY Press, 2005); *Jaffa, the Orange's Clockwork*, directed by Eyal Sivan (Tel Aviv: Trabelsi Productions, 2009), film, 86 min; Mustafa Kabha and Nahum Karlinsky, *The Lost Orchard: The Palestinian-Arab Citrus Industry, 1850-1950* (Syracuse, NY: Syracuse University Press, 2021).

48. Crowfoot and Baldensperger, *From Cedar to Hyssop*, 62.

49. Henry Allen, "The Holy Land Bees," *American Bee Journal* 20, no. 37 (1884): 586.

50. Marmaduke W. Pickthall, *Oriental Encounters: Palestine and Syria, 1894-5-6* (London: W. Colloms Sons and Co., 1918), 11.

51. Baldensperger, *The Immovable East*, xii.

52. Crane, *The World History of Beekeeping and Honey Hunting*, 427. On US equipment utilized by Jewish settlers in Palestine, see "Palestine Honey Production," report sent by the US consul in Jerusalem, Palestine, November 2, 1921, USNA, US State Department, Records of the Department of State relating to the Internal Affairs of Turkey, 1910–1929, film reel 86, 3.

53. "Famous for Honey: An Industry of Palestine in Biblical Days May Be Revived," *Kaskell Free Press*, July 28, 1900. This report was published in a similar version in

various journals and papers around the world, such as "Apiary Interests in Palestine," *Geneva Daily Times*, May 5, 1900; "Honey in Jerusalem," *Bruce Herald*, February 1, 1901 (published in New Zealand); "In the Holy Land," *Hamelitz*, July 5, 1900, 3 (published in Russia in Hebrew).

54. For example, the Lerrer family of Ness Ziona, the first Jewish settler family to adopt the movable frame beehive, complained about theft of hives by neighbors from the Arab village of Mr'ar. The family argued that the (British) government and police were not doing anything to deal with the problem. Lerrer family to the Zionist Commission to the Land of Israel, August 2,1918, Central Zionist Archives [hereafter CZA], L4, 423–1.

55. Harry C. Luke and Edward Keith-Roach, eds., *The Handbook of Palestine, Issued under the Authority of the Government of Palestine* (London: Macmillian and Co., 1922), 165–166.

56. "Loans to Beekeepers," *Davar*, October 25, 1935.

57. Department of Agriculture and Forests, "Sale of Sugar to Beekeepers," *Palestine Post*, October 7, 1934.

58. For details on the establishment of supervision services in regard to the purchase of bees as well as the management of bee diseases, see "Bee Supervision Command," *Davar*, August 26, 1928. According to Israel Robert Blum, the Jewish Beekeepers Association was established in 1929. Israel Robert Blum, "Global Honey Production," CZA, K13, 160, 1, 3. Other documents show that earlier forms of organization existed yet failed. Zionist Management Board of the Zionist Agency, "Proposal to Establish a Cooperative for the Sale of Honey," February 9, 1925, CZA, S25, 521, 6.

59. Breeding experiments with the local bee as well as Caucasian bees were held at the Acre governmental farm in Palestine already in the 1930s, and also by the head of the Jewish agricultural Khadoorie School and beekeeper Natan Fiat. The Italian bee finally gained prominence in Israel in the 1950s.

60. For details about the funding for the Livshutz training with the Baldensperger brothers, see Agricultural Center of the Jewish Workers Organization to the Settlement Department, Re: The work done at farms and settlements, a work diary by Yehoshua Brandstetter, April 24, 1922, CZA, S15, 20915.

61. Press notice by the government, July 25, 1933, CZA, S90, 2121, 1. A similar message was conveyed in the Palestinian Arab newspaper *Filastin* two days later. "Beekeeping in Palestine," *Filastin*, July 27, 1933, 7. For a detailed governmental plan for encouraging Palestinian Arabs to replace clay hives with movable ones, including a loan program, see director of agriculture and forests, Jerusalem, to district commissioners, "Development of Bee-Keeping in Villages," December 22, 1933, ISA, Mem-3, 4243, 1–3.

62. On the colonial tendency to equate Indigenous people with immobility and timelessness, see Johannes Fabian, *Time and the Other: How Anthropology Makes Its Object* (New York: Columbia University Press, 1983).

63. "Expediting the Fellahs," *Davar*, December 14, 1932. On the relation between governance and the standardization of time, see E. P. Thompson, "Time, Work-Discipline, and Industrial Capitalism," *Past and Present* 38 (1967): 56–97. For a discussion of the consolidation of colonial temporality, see On Barak, *On Time: Technology and Temporality in Modern Egypt* (Berkeley: University of California Press, 2013). On time conception and management in the late Ottoman Empire, see Avner Wishnitzer, *Reading Clocks, Alla Turca: Time and Society in the Late Ottoman Empire* (Chicago: Chicago University Press, 2015).

64. "Palestine Honey Production."

65. Baldensperger, "Palestine," 59.

66. Crowfoot and Baldensperger, *From Cedar to Hyssop*, 62.

67. Philip Baldensperger, "Bees in Palestine," *Bee World* 12 (1931): 34–36.

68. "Beekeeping," *Filastin*, March 18, 1922, 1.

69. Israel Robert Blum, "Beekeeping and Its Needs," *Davar*, December 14, 1932. Blum referred here to an article published by Armbruster following his visit to Palestine in 1931. See Ludwig Armbruster, "The Bee in the Orient II: Bible and Bee," *Archiv für Bienenkunde* 7, no. 1 (1932): 1–43. Before immigrating to Palestine, Blum trained with Armbruster in the apiary research institute he established in Dahlem. Armbruster devoted his career to exploring global methods of beekeeping. The Domäne Dahlem farm in Berlin holds a collection of the hives he gathered from around the world as well as photographs indicating that he maintained close professional connections with the Baldensperger family too. With the rise of the Nazi rule, Armbruster, a devoted Christian priest, was forced to resign from his position at the institute due to his support for the Jewish community in Germany. I thank Simon Renkert and Sabine Dankwerts from the Domäne Dahlem archive for their help in obtaining this information.

70. Israel Robert Blum, *The Man and the Bee* (Tel Aviv: Tversky, 1951), 33.

71. "Jewish Beekeepers Conference," *Davar*, August 30, 1939, 4.

72. "Course for Beekeepers," *Davar*, January 10, 1933, 4. Bodenheimer conducted various studies on animal populations. See, for example, Friedrich Simon Bodenheimer, "Studies in Animal Populations II: Seasonal Population—Trends of Honey-Bee," *Quarterly Review of Biology* 12, no. 4 (1937): 406–425. See also Friedrich Simon Bodenheimer, *Animal and Man in Bible Lands* (Leiden: E. J. Brill, 1960). Solomon Leon Skoss (1884–1953), a Philadelphia-based Arabist devoted to Jewish-Arabic studies, was in his early years an enthusiastic beekeeper and traveled to Palestine in 1925. Like Bodenheimer, he published studies relating to the behavior of bees and practice of beekeeping among the ancient Hebrews. Solomon Leon Skoss, *Portrait of a Jewish Scholar: Essays and Addresses* (New York: Block Publishing Company, 1957), 12–13, 137–148.

73. The council was the highest body responsible for the Muslim community's affairs during the British Mandate in Palestine. Haj Amin al-Husseini (1897–1974)

was the mufti of Jerusalem during the British rule—a position that included responsibility over the religious Muslim sites within the city. Al-Husseini was an Arab nationalist and leading figure in the resistance to the growing Jewish settlement and Zionist movement.

74. Israel Robert Blum, "World Congress for Beekeepers," *Davar*, September 25, 1956, 4.

75. "Beekeeping," 1.

76. See Tim Butcher, "Israel No Longer Land of Milk and Honey after 60% Fall in Honey Harvest," *Telegraph*, September 16, 2008, http://www.telegraph.co.uk/earth /earthnews/3351841/Israel-no-longer-Land-of-Milk-and-Honey-after-60-fall-in-honey -harvest.html.

77. This phenomenon—the abrupt disappearance of entire bee colonies—has received growing scientific public attention, especially in North America and Europe. Interestingly enough, the scientific community has often associated this phenomenon with "the Israeli acute paralysis virus (IAPV)," which was named after the nationality of the first researcher who described it. Diana L. Cox-Foster, Sean Conlan, Edward C. Holmes, Gustavo Palacios, Jay D. Evans, Nancy A. Moran, Phenix-Lan Quan, et al., "A Metagenomic Survey of Microbes in Honey Bee Colony Collapse Disorder," *Science* 318 (2007): 283–287.

78. Marwan R. Buheiry, "The Agricultural Exports of Southern Palestine, 1885–1914," *Journal of Palestine Studies* 10, no. 4 (1981): 67.

79. Bruce Bower, "Excavators Find Honey of a Discovery: Israeli Site Yields Oldest Known Example of Beekeeping," *Science News*, September 27, 2008, 11.

80. Amihai Mazar and Nava Panitz-Cohen, "It Is the Land of Honey: Beekeeping at Tel Reḥov," *Near Eastern Archaeology* 70, no. 4 (2007): 218, 213–214.

CHAPTER 2

1. Petition of goat owners of A'ara village to the prime minister of Israel, December 2, 1952, Israel State Archives [hereafter ISA], Gimel Lamed-19, 17022.

2. This petition, as all others I found in the Israeli State Archives and the Israeli Defense Force's Archives were written in Arabic (either by hand or typed), and then translated by the governing rule to either English or Hebrew. The 1948 War shifted power structures and population composition in Palestine, of which major parts became the State of Israel. During the war, many hundreds of thousands of Palestinian Arabs forced into exile (most estimations vary from 700,000 to 900,000 people), and only a small few were able to stay and lived under the new military rule (for Palestinians only), which lasted until 1966. Palestinians today comprise approximately 21 percent of the citizens of the State of Israel.

3. According to British accounts, there were 571,289 goats over one year of age in Palestine in 1926, 307,316 in 1937, and approximately 325,000 in 1943 (that

is in comparison to 290,854 sheep in 1926, 177,838 in 1937, and approximately 244,000 in 1943); the first full livestock census took place in 1930. Roza El-Eini, *Mandated Landscape: British Imperial Rule in Palestine, 1929–1948* (New York: Routledge, 2006), 226. On November 21, 1918, as part of her impressions as a participant in the American Red Cross Commission to Palestine, US nurse Edith Madeira wrote about the unavailability of dairy cows and prevalence of goats: "Cows being out of the question we are trying to buy goats for milk for the babies. They will have a goat herd and come into the hospital yard and be milked there twice a day. Doesn't that sound queer and oriental?" Historical Society of Philadelphia Archives, Edith Madeira Papers, 3.

4. On the use of the law as a tool for seizing control as well as the consequential resistance (for both the colonizer and colonized) in British Palestine, see Geremy Forman and Alexander Kedar, "Colonialism, Colonization and Land Law in Mandate Palestine: The Zor al-Zarqa and Barrat Qisarya Land Disputes in Historical Perspective," *Theoretical Inquiries in Law* 4, no. 2 (2003): 491–540.

5. There are a few differences in this regard between the British rule and Jewish settlers, as afforestation efforts by the British had much more to do with taming nature rather than reviving it. As G. V. Jacks and R. O. Whyte wrote in *The Rape of the Earth*, "Erosion is a modern symptom of maladjustment between human society and its environment. It is a warning that Nature is in full revolt against the sudden incursion of an exotic civilization into her ordered domains. Men are permitted to dominate Nature on precisely the same condition as trees and plants, namely on conditions that they improve the soil and leave it a little better for their posterity than they found it." G. V. Jacks and R. O. Whyte, *The Rape of the Earth: A World Survey of Soil Erosion* (London: Faber and Faber Ltd., 1939), 26.

6. See, for example, Timothy Mitchell, *Rule of Experts: Egypt, Techno-Politics, Modernity* (Berkeley: University of California Press, 2002); Diana K. Davis, *Resurrecting the Granary of Rome: Environmental History and French Colonialism Expansion in North Africa* (Athens: Ohio University Press, 2007); Priya Satia, *Spies in Arabia: The Great War and the Cultural Foundations of Britain's Covert Empire in the Middle East* (Oxford: Oxford University Press, 2008); Diana K. Davis and Edmund Burke III, eds., *Environmental Imaginaries of the Middle East and North Africa* (Athens: Ohio University Press, 2011); On Barak, *On Time: Technology and Temporality in Modern Egypt* (Berkeley: University of California Press, 2013).

7. For the importance of agricultural improvement and later development to the British Empire, see Richard Drayton, *Nature's Government: Science, Imperial Britain and the "Improvement" of the World* (New Haven, CT: Yale University Press, 2000); Joseph Morgan Hodge, *Triumph of the Expert: Agrarian Doctrines of Development and the Legacies of British Colonialism* (Athens: Ohio University Press, 2007). Gregory A. Barton argues that the British "made forestry the most important aspect of agricultural development in the Middle East after the Second World War." Gregory A. Barton, "Environmentalism, Development and British Policy in the Middle East 1945–56,"

Journal of Imperial and Commonwealth History 38, no. 4 (2010): 619–639. For afforestation policies in Palestine, see Roza El-Eini, "British Forestry Policy in Mandate Palestine, 1929–48," *Middle Eastern Studies* 35, no. 3 (1999): 72–155; Gideon Biger and Nili Lifshitz, "The Afforestation Policy of the British Government in the Land of Israel," *Ofakim Begeographia* 40–41 (1994): 5–16.

8. A great wealth of historical work has focused on forestry in the British Empire and colonial India in particular. See, for example, Peder Anker, *Imperial Ecology: Environmental Order in the British Empire, 1895–1945* (Cambridge, MA: Harvard University Press, 2001); David Arnold, *The Tropics and the Traveling Gaze: India, Landscape and Science, 1800–1856* (Seattle: University of Washington Press, 2006). For a review essay on the literature, see Kalyanakrishnan Sivaramakrishnan, "Science, Environment and Empire History: Comparative Perspectives from Forests in Colonial India," *Environment and History* 14 (2008): 41–65. On forestry and ecology in French colonial Maghreb, see Davis, *Resurrecting the Granary of Rome*.

9. For information on the Jewish National Fund, role of afforestation in Zionist symbolism, and expansion of Jewish settlement in the West Bank, see Irus Braverman, *Planted Flags: Trees, Land, and Law in Israel/Palestine* (Cambridge: Cambridge University Press, 2009). Braverman details various reasons for why the pine tree, the "Jewish tree," became central to the planting projects of both the British government and Jewish National Fund: the pine tree grows quickly and helps construct a "European-style" landscape. Additionally, pine is defined as a forest and not a fruit tree, which is thereby considered a form of noncultivation and has been used (under Article 78 of the Ottoman land code that is still in force) for declaring various lands as state land. Irus Braverman, "The Tree Is the Enemy Soldier: A Sociolegal Making of War Landscapes in the Occupied West Bank," *Law and Society Review* 42, no. 3 (2008): 456, 462–463; Irus Braverman, "Planting the Promised Landscape: Zionism, Nature, and Resistance in Israel/Palestine," *Natural Resources Journal* 49 (2009): 343. For other works dealing with the Zionist planting fervor and great afforestation project, see Shaul E. Cohen, *The Politics of Planting: Israeli-Palestinian Competition for Control of Land in the Jerusalem Periphery* (Chicago: University of Chicago Press, 1993); Yael Zerubavel, *Recovered Roots: Collective Memory and the Making of Israeli National Tradition* (Chicago: University of Chicago Press, 1995); Alon Tal, "Combating Desertification: Evolving Perceptions and Strategies," in *Between Ruin and Restoration: An Environmental History of Israel*, ed. Daniel Orenstein, Alon Tal, and Char Miller (Pittsburgh: University of Pittsburgh Press, 2013), 119–122.

10. Sgd. G. N. Sale, conservator of forests, "Memorandum on Control of Grazing," attached to a letter from Sale to F. R. Mason, director of agriculture and fisheries, March 22, 1943, ISA, Mem-13, 5109. In *Resurrecting the Granary of Rome*, Davis analyzes similar claims regarding desertification processes in North Africa under French colonial rule. This paradigm of decline was also applied to the entirety of the Middle East throughout the twentieth century. As late as 1971, a US agricultural development expert wrote, "Today's traveler finds it almost impossible to believe that most

of the now barren slopes and mountains with annual precipitation in excess of 12 inches were once forested. . . . The influence of uncontrolled grazing is also evident. The ground cover of the depleted forests is now mainly composed of thorny, unpalatable tragacanth species. . . . Full control of the land and the animals must be in the hands of the technicians." C. Kenneth Pearse, "Grazing in the Middle East: Past, Present, and Future," *Journal of Range Management* 24, no. 1 (1971): 13, 16.

11. Sgd. M. C. Alhassid to director, "Grazing Grounds—Management of, Copy" January 4, 1941, 2, ISA, Mem-19, 20.

12. Sale, "Memorandum on Control of Grazing."

13. Sgd. F. R. Mason, acting chair, Soil Conservation Board, to district commissioners, "Allocation of State Domains as Grazing Grounds," September 10, 1946, ISA, Mem-29, 4982.

14. Major C. S. Jarvis, "The Arab, the Goat, and the Camel: Destroyers of the Desert," *Palestine Post*, October 11, 1934. "By eating out the heart of every living plant," he said, "they had removed all the binding material provided by Nature."

15. Sale, "Memorandum on Control of Grazing."

16. Jacks and Whyte, *The Rape of the Earth*, 42.

17. Timothy Mitchell discusses the gradual move from wood to coal and then to oil as the major source for global energy. As part of this change, forests lost their importance to the global economy and political structure from the mid-nineteenth century onward. Timothy Mitchell, *Carbon Democracy: Political Power in the Age of Oil* (London: Verso, 2011), 15.

18. Conservator of forests to district commissioner, Jerusalem, "Suggested Grazing in Allenby Bridge State Forests," March 11, 1941, ISA, Mem-13, 5109.

19. Tal, "Combating Desertification," 120.

20. On the relation between forest management and the control of the state, see James C. Scott, *Seeing Like a State: How Certain Schemes to Improve the Human Condition Have Failed* (New Haven, CT: Yale University Press, 1998). For a discussion of the emergence of subjectivity in relation to colonial forest regulations, see Arun Agrawal, *Environmentality: Technologies of Government and the Making of Subjects* (Durham, NC: Duke University Press, 2005).

21. Sgd. F. H. Taylor, district engineer of the Lydda, to chair of the Soil Conservation Board, "Grazing," July 11, 1942, ISA, Mem-13, 5109. Mitchell (*Carbon Democracy*, 21) calls such interference with the movement of resources initiated by imperial powers "sabotage." On the expansion of the railway system in British Palestine, see David Tirosh, *The Emek Train* (Tel Aviv: Society for the Protection of Nature [HaHevra LeHaganat HaTeva], 1988); Zachary Lockman, *Comrades and Enemies: Arab and Jewish Workers in Palestine, 1906–1948* (Berkeley: University of California Press, 1996).

22. For the economic basis of the 1936 events, see Mahmoud Yazbak, "From Poverty to Revolt: Economic Factors in the Outbreak of the 1936 Rebellion in Palestine," *Middle Eastern Studies* 36, no. 3 (2000): 93–113. Yazbak's arguments defy other works contending that a peasant's economic status was improving in the 1930s. See, for example, Yuval Arnon-Ohana, *Peasants in the Arab Revolt in the Land of Israel, 1936–1939* (Tel Aviv: Shiloah Institute, Tel Aviv University, 1978), 36. Peasants had a major role in the 1936–1939 revolts, and some acts of rebellion were targeted at centers of state-supported agricultural research and education. See, for example, the case of the fire at the Arab Khadoorie Agricultural School at Tulkarm discussed in folder, ISA, Mem-8, 4304.

23. Matthew Hughes, "From Law and Order to Pacification: Britain's Suppression of the Arab Revolt in Palestine, 1936–39," *Journal of Palestine Studies* 39, no. 2 (2010): 6–22. While stressing the element of surprise vis-à-vis the Arab Revolt, it is important to note how "social unrest and strikes erupted throughout the colonial empire in the late 1930s." Hodge, *Triumph of the Expert*, 13.

24. Rashid Khalidi, *The Iron Cage: The Story of the Palestinian Struggle for Statehood* (Boston: Beacon Press, 2006). It is widely agreed that the events of 1936, and more broadly the Arab Revolt of 1936–1939, were transformative in the relation between Arabs and Jews as well as the relation between Palestinian Arabs and Jews to the British government. In his recent book, however, Hillel Cohen identifies 1929 as the "year zero" of the Jewish-Arab conflict. He argues that the violent acts between Jews and Arabs solidified a new binary understanding of these populations. Hillel Cohen, *1929: Year Zero of the Jewish-Arab Conflict* (Jerusalem: Keter, 2013).

25. Sale, "Afforestation and Soil Conservation," 3.

26. Sale, "Memorandum on Control of Grazing."

27. Mason noted that "we must endeavour to ascertain the true number of animals now existing in the country." Quoted in Sale, "Memorandum on Control of Grazing." Enumeration practices were not new to Palestine, but in this case, their scope was novel. Organized censuses of people and livestock became prevalent in the last years of the Ottoman rule. For a discussion of the Animal Enumeration Law of 1905, see El-Eini *Mandated Landscape*, 226.

28. Chief veterinary officer, "Goat Raising as a Paying Proposition" (extract from "Sheep and Goats in Palestine by Dr. S. Hirsh Published in the *Bulletin of the Palestine Economic Society* [February 1933])," January 14, 1941, ISA, Mem-13, 5109.

29. The tendency to undermine the logic behind behaviors that appear not to align with a particular understanding of improvement and progress is central to the discourse of Western rationality, for which statistics became a tool of confirmation. For a similar case in which official improvement efforts delegitimized the logic of commonly used agricultural practices, see Deborah Fitzgerald, *Every Farm a Factory: The Industrial Ideal in American Agriculture* (New Haven, CT: Yale University Press, 2003).

30. Sgd. G. N. Sale, conservator of forests, "Note," October 13, 1939, ISA, Mem-13, 5109. On the changes in animal taxation from the Ottoman system to the British Mandate, see Roza El-Eini, "Governmental Fiscal Policy in Mandatory Palestine in the 1930s," *Middle Eastern Studies* 33, no. 3 (1997): 570–596.

31. For a detailed analysis of the gradual rise in taxes in the last years of the Ottoman rule in Palestine and first half of the British rule, and its affects on the peasant population as well as urbanization processes, see Yazbak, "From Poverty to Revolt."

32. Sale, "Memorandum on Control of Grazing."

33. District commissioner of Samaria to district commissioner of Galilee, "Paragraph 2(a) of the Minutes of the Meeting of the District Commissioner's Conference Held on 30th June, 1944," July 20, 1944, ISA, Mem-13, 5109.

34. A. Y. Goor, conservator of forests, Department of Forests, to director of agriculture and fisheries, letter, October 20, 1946, ISA, Mem-13, 5109.

35. Goor, Department of Forests, to director of agriculture and fisheries.

36. Sale, "Memorandum on Control of Grazing."

37. D. C. MacGillivray to district commissioners, "1946 Shepherds (Licensing) Ordinance," November 13, 1946, ISA, Mem-10, 22. According to the ordinance, "Each shepherd licensed under the Ordinance will be issued with a small metal identity disc which he will be required to carry while grazing his flock," D. C. MacGillivray to district commissioners, letter, April 5, 1947, ISA, Mem-13, 5109. In addition to shepherds, goats too had to carry a tag. On the practice of tagging goats, see acting district commissioner of Galilee district to chief secretary, April 10, 1947, ISA, Mem-1, 4190.

38. Minutes of the meeting of the Committee for the Preservation of Trees and Prohibition of Grazing, July 17, 1942, 3, ISA, Mem-13, 5109.

39. "Grazing (Control) Ordinance: An Ordinance to Control the Grazing of Domestic Animals—Revised Copy," n.d., 2, ISA, Mem-13, 5109.

40. Sale, "Note."

41. "Of Their Methods for Destroying the Resources of Arabs!," *Filastin*, January 29, 1947, 2.

42. "Of Their Methods for Destroying the Resources of Arabs!" See also "The Project of Pinching Holes in Goat Ears and the Sale of Lands," *Filastin*, January 31, 1947, 4.

43. The scientific understanding of the land and its creatures has often been contested under colonial rules. For a discussion of the interwar debates regarding land management in Africa, see Helen Tilly, *Africa as a Living Laboratory: Empire, Development, and the Problem of Scientific Knowledge, 1870–1950* (Chicago: University of Chicago Press, 2011). Hodge (*Triumph of the Expert*, 7–9) similarly stresses that colonial development policies were constantly being shaped by debates, divisions, and doubts between scientific experts.

44. Foresters were aware of the challenges from within: "It is generally believed that the Veterinary section of your department raised objection to the reduction of the number of free ranging goats in this country." Goor, Department of Forests, to director of agriculture and fisheries.

45. G. S. Emanuel, chief veterinary officer to director of department of agriculture and fisheries, letter, July 19, 1943, ISA, Mem-13, 5109.

46. G. C. L. Bertram, chief fisheries officer, to conservator of forests, "The Goat and the Man," November 13, 1945, ISA, Mem-13, 5109; For a discussion of various objections to Sale's plans of afforestation, see El-Eini, "British Forestry Policy in Mandate Palestine."

47. G. C. L. Bertram, chief fisheries officer, "The Best Domestic Animal for Poor Peasantry," November 13, 1945, ISA, Mem-13, 5109.

48. Bertram, "The Best Domestic Animal for Poor Peasantry."

49. Senior veterinary officer to chief veterinary officer, "Grazing Control Ordinance—Comments," June 23, 1943, ISA, Mem-13, 5109. See also sgd. W. R. McGeagn, district commissioner of Jerusalem, to conservator of forests, "Grazing in Closed Areas between Allenby Bridge and Maghtas," March 6, 1941, ISA, Mem-13, 5109.

50. For a discussion of the pervasiveness of criticism generated from within the community of colonial experts, see Tilley, *Africa as a Living Laboratory*, 23–25.

51. District commissioner of Galilee to chief secretary, letter (secret), July 4, 1944, ISA, Mem-26, 4304.

52. Chief veterinary officer, "Goat Raising as a Paying Proposition" (extract from a Paper on "Sheep and Goats in Palestine' by Dr. S. Hirsh Published in the *Bulletin of the Palestine Economic Society* [February 1933])," January 14, 1941, ISA, Mem-13, 5109.

53. According to British experts, "All attempts to introduce a scheme to reduce the number of goats has so far failed." El-Eini, *Mandated Landscape*, 228. On the various problems with counting people in British Palestine and the State of Israel, see Anat Leibler, "Statisticians' Ambition: Governmentality, Modernity and National Legibility," *Israel Studies* 9, no. 2 (2004): 139–140.

54. "Minutes of the 139th Meeting of the First Parliament, May 15, 1950: C. Plant Protection Law, 1950 (First Call)," in *Parliament Minutes, Second Sitting of the First Israeli Parliament* (Jerusalem: Israeli Government Press, 1950), 1368–1369, ISA. I thank Tzach Glasser for the various documents regarding the 1950 Plant Protection Law and introducing me to the goats he works with.

55. On the changing understanding of property and landownership in Palestine in the late nineteenth and early twentieth centuries, see the introduction and chapter 3. Much research has been devoted to the friction between changing concepts of land and landownership. Several environmental historians have paid attention to the ways people in settler societies versus Indigenous people understood their

property. For example, for a discussion of the process of fencing, see William Cronon, *Changes in the Land: Indians, Colonists, and the Ecology of New England* (New York: Hill and Wang, 1983). Few, however, chose animals as a way to demonstrate such conflict. For example, for a discussion of cows in early colonial America and the challenges they posed as they crossed fences, see Virginia D. Anderson, *Creatures of Empire: How Domestic Animals Transformed Early America* (Oxford: Oxford University Press, 2004).

56. "Minutes of the 139th Meeting of the First Parliament."

57. Yohanan Bader (Herut movement) quoted in "Minutes of the 139th Meeting of the First Parliament."

58. Eliyahu Moshe Ganhovsky (from the United Religious Front) quoted in "Minutes of the 139th Meeting of the First Parliament."

59. According to British registrations, there were 250,000 goats in Palestine in 1946. An experiment that involved the tagging of goats by the Department of Forestry in the Bethlehem and Jenin areas demonstrated that taxes were only paid for 35 percent of the goats there. From that study, the conservator of forests concluded that the actual number of goats was three times the amount declared by owners, hence reaching the number of 750,000. Mr. A. V. Goor, conservator of forests, "Enumeration of Goats," minutes from a Department of Conservation conference, Jerusalem, September 27, 1946, ISA, Mem-1, 4190. Similarly, the Israeli government managed to register 65,000 goats owned by Arabs in Israel in 1950, but officials in the Ministry of Agriculture estimated that their actual number was closer to 100,000. Y. Kotzer, director of the Arab Section, to Mr. Ben-David, general secretary in the Ministry of Agriculture, "Termination of Goat Herds in the Arab Villages," November 23, 1950, ISA, Gimel Lamed-19, 17022; Dr. M. Pintchi, director of the Sheep and Goats Section, and Y. Kotzer, director of the Arab Village Section, Ministry of Agriculture, to the general secretary of the Ministry of Agriculture, "A Plan to Execute the Law for the Termination of Goats," February 8, 1951, ISA, Gimel Lamed-19, 17022. Moreover, it is hard to determine the fate of Palestinian goats during the war; it was extremely unclear as to how many of them were killed, how many were left with their owners, and how many remained in the territory that was not part of the State of Israel. Some sources indicate that many Palestinians took their livestock with them. Donald Stevenson, a member of the Quaker American Friends Service Committee who took part in relief work in Gaza after the 1948 war, recounted a conversation with Gordon Clapp, chair of the US Tennessee Valley Authority who headed an economic survey mission under a Palestine Conciliation Commission initiated by the United Nations: "I asked Clapp what was going to be done about the 'black scourge of the East,' the ubiquitous goat." Clapp answered, "Israel has solved this problem by running most of the Arabs and their goats out of its territory." Quoted in Benjamin N. Schiff, *Refugees unto the Third Generation: UN Aid to Palestinians* (Syracuse, NY: Syracuse University Press, 1995), 20.

60. Quoted in Schiff, *Refugees unto the Third Generation*.

61. As a result of this new understanding of the relation between sheep and goats, instructors hired by the Ministry of Agriculture told farmers to separate the sheep from the goats. See "No Progress with Mixture: The Racialization of Goats" section.

62. Mr. M. Erem, chair, "First Discussion about the Plant Protection (Damage from Goats) Law," protocol of the Committee of Economics of the (first) Israeli Parliament, May 23, 1950, ISA, Kaf-1, 22.

63. S. Goren, "First Discussion about the Plant Protection (Damage from Goats) Law"; A. Hushi, "First Discussion about the Plant Protection (Damage from Goats) Law."

64. Mr. Bader, chair, "Final Discussion about the Plant Protection (Damage from Goats) Law," protocol of the Committee of Economics of the Israeli Parliament, June 20, 1950, ISA, Kaf-1, 22.

65. Sh. Lavie, "Final Discussion about the Plant Protection (Damage from Goats) Law."

66. Mr. Halperin, "Final Discussion about the Plant Protection (Damage from Goats) Law."

67. Pintchi and Kotzer to the general secretary.

68. This idea of replacement was discussed among foresters during the British rule, yet it never progressed to a full-fledged plan of action. See El-Eini, *Mandated Landscape*, 226. For a discussion regarding an old goat owner in Kababir who claimed he was unable to replace his goats with cows, see Forest inspector of Haifa district to conservator of forests, "Grazing," August 30, 1945, ISA, M-43, 4306.

69. Pintchi and Kotzer to the general secretary. For similar details of the plan, see Y. Kotzer, director of the Arab Village Section, to A. Ben-David, general secretary, Ministry of Agriculture, "Termination of Herds of Goats in Arab Villages" November 23, 1950, ISA, Gimel Lamed-19, 17022.

70. M. Karniel and Co., Kfar Tabor, to Mr. Hanuki, Division of Agricultural Development, letter, March 14, 1954, ISA, Gimel-5, 2868.

71. Arab Village Section, Ministry of Agriculture, to the military governor of Kefar Yona, "Execution of the Law for the Termination of Goats," June 13, 1951, ISA, Gimel Lamed-19, 17022.

72. A. Hanuki, Ministry of Agriculture, to the adviser for Arab matters in the prime minister's office, "Termination of the Herds of Goats," January 14, 1953, ISA, Gimel Lamed-19, 17022.

73. The "Triangle Area" refers to a concentration of Palestinian villages and towns that became part of the territory of the State of Israel in 1948, and was under military rule until 1966. "List of Goats and Sheep in the Arab Villages in the Triangle Area," February 24, 1954, ISA, Gimel-5, 2868.

74. Mr. A. Hanuki, Division of Agricultural Development, to Mr. Sh. Zamir, officer of the development of the triangle, Ministry of Agriculture, letter, March 1954, ISA, Gimel-5, 2868.

75. James C. Scott, *Weapons of the Weak: Everyday Forms of Peasant Resistance* (New Haven, CT: Yale University Press, 1985). As for the goats themselves, we might think of everyday forms of animal resistance. See Jason Hribal, *Fear of the Animal Planet: The Hidden History of Animal Resistance* (Chico, CA: AK Press, 2010).

76. Much of the research devoted to common people and peasant life under the Ottoman rule is based on petitions sent from around the empire to the central government in Istanbul. Amy Singer, Beshara Doumani, and Yuval Ben-Bassat, for example, examine petitions to Ottoman officials as a way to analyze the lives of peasants in Palestine. See Amy Singer, *Palestinian Peasants and Ottoman Officials: Rural Administration around Sixteenth-Century Jerusalem* (Cambridge: Cambridge University Press, 1994); Beshara Doumani, *Rediscovering Palestine: Merchants and Peasants in Jabal Nablus, 1700–1900* (Berkeley: University of California Press, 1995); Yuval Ben-Bassat, *Petitioning the Sultan: Justice and Protest in Late Ottoman Palestine* (London: I. B. Tauris, 2013).

77. Assaf Likhovski, "Between 'Mandate' and 'State': Re-Thinking the Periodization of Israeli Legal History," *Journal of Israeli History* 19, no. 2 (1998): 39–68; David Schorr, "A Prolonged Recessional: The Continuing Influence of British Rule on Israeli Environmental Law," in *Between Ruin and Restoration: An Environmental History of Israel*, ed. Daniel Orenstein, Alon Tal, and Char Miller (Pittsburgh: University of Pittsburgh Press, 2013), 209–228.

78. According to Scott's analysis, this happens when the violation of the moral economy of the subsistence ethics is great enough. Scott has written extensively about peasants' economy and means of resistance to the governing power, mostly focusing on Southeast Asia. His interest lies in those times when peasants, in spite of great risks, choose to revolt. Assuming that peasants have limited power vis-à-vis state structures, Scott does not focus on peasants' attempts to shape structural changes with tools of the state. James C. Scott, *The Moral Economy of the Peasant: Rebellion and Subsistence in Southeast Asia* (New Haven, CT: Yale University Press, 1976); Scott, *Weapons of the Weak*; James C. Scott, *The Art of Not Being Governed: An Anarchist History of Upland Southeast Asia* (New Haven, CT: Yale University Press, 2009).

79. Here I employ Natali Zemon Davis's insight on the creativity demonstrated by common people who address officials of the state. Analyzing letters of remission produced in sixteenth-century France, Davis looks at fictitious elements in the crafting of a narrative for the purpose of mitigating criminal sentences. Natali Zemon Davis, *Fiction in the Archives: Pardon Tales and Their Tellers in Sixteenth-Century France* (Stanford, CA: Stanford University Press, 1987).

80. Mukhtars of the tribe Esh Shibli to the high commissioner of Jerusalem, petition, January 16, 1946, ISA, Mem-77, 313. Mukhtars are the appointed heads of Palestinian villages. Dunam (or dunum) is a common Ottoman unit of land area, which has been in use in Palestine/Israel ever since.

81. Mukhtars to high commissioner.

82. Muhamad Daud in the name of the goat owners of 'Ein-Sahala, the Triangle Area, to the chair of Parliament, petition, December 10, 1952, ISA, Gimel Lamed-19, 17022.

83. Goat owners of Sakhnin village to the minister of agriculture, petition, July 21, 1954, ISA, Gimel-Lamed-19, 17022. Pruta was the domination of currency used in Israel during 1948–1960, with its name borrowed from Mishnaic Hebrew.

84. Not only Palestinian, but some Jewish Israelis complained about shrinking grazing areas. The people of Kibbutz Lehavot, for example, wrote to the Jewish National Fund (KKL) about the problem of grazing areas needed for their sheep herd in 1955. Lehavot to Mr. Ra'anan Weitz, KKL, "Plowing Grazing Land," August 17, 1955, Central Zionist Archives [hereafter CZA], KKL5, 22311.

85. "A Severe Dispute between the Druze and the Ministry of Agriculture regarding the Grazing of Goats," *Ha'aretz*, January 24, 1954, ISA, Gimel Lamed-19, 17022.

86. Goat owners in the villages of the Nazareth area to the chair of Parliament, petition, December 22, 1952, ISA, Gimel Lamed-19, 17022.

87. Hassan Taka Muhamad, Sakhnin village, in the name of the people of the Sakhnin village, to the prime minister, "The Plant Protection (Damage of Goat) Law 1950," December 28, 1952, ISA, Gimel Lamed-19, 17022.

88. Owners of goats in the Baqa al-Gharbiyye area to the minister of agriculture, petition, November 21, 1952, Israel Defense Forces Archives, 1954-300-66, 84.

89. Note from unknown writer, circa 1952, ISA, Gimel Lamed-19, 17022.

90. Yitzhak Shani, department manager, military rule—central region, to Mr. D. Zackariah, Ministry of Agriculture, "Re: The Request of the Goat Owners in the Baqa al-Gharbiyye Area," December 1, 1952, Israel Defense Forces Archives, 1954-300-66, 83.

91. The Goats Section of the Hebrew Shepherds Association was established in 1941. H. Helperin to Sh. Shimerman, Kfar-Melel, November 19, 1940, Lavon Institute for Labor Research Archives [hereafter LILR], IV-235-1-737; "Purchase of Sheep and Goats: Importation according to Countries, 1951–1958," CZA, S, 15, 40983.

92. Hebrew Shepherds Association, Goats Section, newsletter 2, February 4, 1945, 1, ISA, Mem-20, 653.

93. Yonatan Amir, chair of Aziza, to David Khahana, Agricultural Center, "A Memorandum on the State of Raising White Goats in the Country," July 22, 1965, LILR, file IV-235–5, 3871.

94. Organization of the Hebrew Shepherds, Goats Section, newsletter [no number], 1941, 3, ISA, M-20, 653. Experts detailed what a good, rational stall was: "A) where the food within it always stays clean and the goat does not have an opportunity of spoiling it; B) when it is organized in a matter in which the access to the food would challenge the goat, since the goat doesn't eat when it [female gender] is too comfortable. . . . [M]easuring the mild and testing: the question of the quality of the goat and the volume of their yield is a main question in the worthwhileness of holding it. Only goats that produce plenty are appropriate and worth caring for. The quality of goats should not be determined by estimation or memory. Only exact and meticulous measurement of milk ensures setting the value of the goats. The measurements of milk should be registered in a particular uniform order that ensures the comparison of the yields." M. Schorr, "Instructions for Goat Growers with the Coming of Summer," May 8, 1946, ISA, Mem-20, 653.

95. Schorr, "Instructions for Goat Growers with the Coming of Summer." The Latin origin of the word "capricious" is *capra*—goat.

96. Hebrew Shepherds Association, Goat Growers Section, newsletter 8, November 10, 1942, 2, LILR, IV-235-1-737. This argument is also prevalent among contemporary Jewish goat owners, as I discovered in conversations held in July 2012. In fact, to quote from Dov Beker's "Instructions for the Season," "It is now the grazing season. . . . [I]t is not advisable to gather the goats into a herd, as is customary in bigger farms. But this view does not object taking 5–6 goats together to graze. The best method is: carrying the goat with a rope and then it is possible to take advantage of the richest places of high, soft and tasty grass." Thus the best way for goats to graze was for them to be tied with a rope to a post. Hebrew Shepherds Association, newsletter 2.

97. Dov Beker, *Sheep and Goat Rearing* (Ein-Harod: Hebrew Shepherds Association, 1948), 68.

98. Ag. Moshe Schorr, *The House Goat* (Tel Aviv: Hakarmel [published by the chief supervisor of agricultural education, Ministry of Education and Culture], 1949), 4.

99. "Suggestion of Law 47 by the Government of Israel: Plant Protection (Damage of Goats) Law," June 19, 1950, ISA, Gimel-23, 5400.

100. "Protocol of the Hebrew Shepherds Association," November 21, 1950, LILR, 280 IV, 1.

101. Ann Laura Stoler, *Carnal Knowledge and Imperial Power: Race and the Intimate in Colonial Rule* (Berkeley: University of California Press, 2010), 79–111. For an additional layer to the fear of mixture and how it related to the bodies of animals and their breeding, see chapter 4.

102. G. Sharoni, "The Goats of Israel Assembled in the College," *Ma'ariv*, August 30, 1953, 2.

103. "Protocol of the Hebrew Shepherds Association."

104. Haim Schwartz, regional instructor for sheep and goats raising, to the Training Department at the Agricultural Research Station of Rehovot of the Jewish Agency, "Report no. 2b on the Situation of Sheep and Goats in the Ashkelon-Majdal Area," May 1955, Hashomer Hatzair Archives, Yad-Ya'ari, Givat Haviva, 125, 3, 4.

105. Tawfiq Canaan, *Mohammedan Saints and Sanctuaries in Palestine* (London: Luzac and Co., 1927), 243–244 (reprinted from *Journal of the Palestine Oriental Society*). On the nativist ethnography of Canaan and his Jerusalem circle, see Salim Tamari, *Mountain against the Sea: Essays on Palestinian Society and Culture* (Berkeley: University of California Press, 2009), 93–112.

106. Albert Boime, "William Holman Hunt's 'The Scapegoat': Rite of Forgiveness / Transference of Blame," *Art Bulletin* 84, no. 1 (2002): 94–114. The representation of Jesus as a harmless lamb is a common theme in religious art of all types; the nontraditional choice of a goat, and a rather miserably looking one, was the reason for such public fury.

107. Sale, "Afforestation and Soil Conservation."

108. Anthony S. Wohl, "'Ben JuJu': Representations of Disraeli's Jewishness in the Victorian Political Cartoon," *Jewish History* 10, no. 2 (1996): 89–134.

109. Sholom Aleichem, *The Bewitched Tailor* (Moscow: Foreign Languages Publication House, 1960); Isaac Bashevis Singer, *Zlateh the Goat and Other Stories* (New York: HarperCollins, 1966).

110. See, for example Marc Chagall, *I and the Village*, 1911, http://www.moma.org /collection/object.php?object_id=78984; Reuven Rubin, *Self-Portrait with a Goat*, 1924, http://www.bridgemanimages.com/en-GB/asset/825853/rubin-reuven-1893-1974 /self-portrait-1924-oil-on-canvas-on-masonite.

111. Aharon Pri'el, "'A Green Patrol' Will Operate against Bedouin Herds," *Ma'ariv*, August 4, 1976, 6. According to historian and activist Gadi Algazi, the Green Patrol, one of the famous policing forces in Israel, perpetually violates citizen rights. Gadi Algazi, "From Gir Forest to Um-Hiran: Comment on Colonial Nature and Its Conservators," *Teoria Vebikoret* 37 (2010): 232–253.

112. Yosef Algazi, "Ransoming Captive Camels," *Ha'aretz*, October 14, 1994; Dror Shohet and Dan Shohet, "Our Camels Are Part of the Desert," Facebook, 2012, https://www.facebook.com/notes/dror-shohet/%D7%94%D7%92%D7%9E%D7% 9C%D7%99%D7%9D-%D7%A9%D7%9C%D7%A0%D7%95-%D7%94%D7%9D- %D7%97%D7%9C%D7%A7-%D7%9E%D7%94%D7%9E%D7%93%D7%91%D7 %A8/10151891597981866. The Ministry of Agriculture began to systematically tag camels in 2015, and institutionalized their surveillance in 2018 with new regulations requiring the implantation of identification chips. See Almog Ben Zikri, "Only a Third of Israeli Camels Have ID Chips—and It Could Be Deadly," *Ha'aretz*, October 7, 2019, https://www.haaretz.com/israel-news/.premium-only-a-third-of-bedouin-cam els-in-israel-s-south-have-id-chips-1.7949607.

113. Avi Perevolotsky and No'am G. Seligman, "Role of Grazing in Mediterranean Rangeland Ecosystems," *Bioscience* 48 (1998): 1007–1017; Imanuel Noy-Meir and Talya Oron, "Effects of Grazing on Geophytes in Mediterranean Vegetation," *Journal of Vegetation Science* 12, no. 6 (2001): 749–760; Elizabeth Wachs and Alon Tal, "Herd No More: Livestock Husbandry Policies and the Environment in Israel," *Journal of Environmental Ethics* 22 (2009): 401–422. Davis (*Resurrecting the Granary of Rome*, 185–186) describes a similar shift in Algerian environmental policy.

114. Interview with Dorit Kababia, manager of the Sheep and Goats Section, Ministry of Agriculture, Beit Dagan, September 15, 2013, regarding the efforts and failures to convince Bedouin shepherds to herd in the mountains of the Jerusalem area.

115. Records of the Book of Laws 2717, Ministry of Justice, Israel, May 13, 2018, 662.

CHAPTER 3

1. A kibbutz is a communal agricultural settlement. The name of the kibbutz, Merhavia (which was also the name of a nearby moshav, or a semicooperative settlement), has biblical origins and roughly translates into "the space of God."

2. Letter from Professor Felix G. Sulman to Aharon Harari, August 10, 1962, Kibbutz Merhavia Archives [hereafter KMA], Aharon Harari Collection, 2.7.

3. "David Zamir," biographical notes, KMA, David Zamir Collection; Isar Lavi, "'The Troubadour' of Sheep Management," in *David Zamir (A Collection for His Memory)* (Merhavia: Shepherds Association and Kibbutz Merhavia, 1968), 13.

4. On the importance of the study of pastoralism for the environmental history of the Middle East, see Edmund Burke III, "Pastoralism and the Mediterranean Environment," *International Journal of Middle East Studies* (2010): 663–665.

5. For more on ideas of the collective ownership of land, *mushāʿ* land system, and ways in which older forms of land management changed during the late Ottoman rule and British Mandate, see Amos Nadan, "Colonial Misunderstanding of an Efficient Peasant Institution: Land Settlement and Mushāʿ Tenure in Mandate Palestine, 1921–47," *Journal of Economic and Social History of the Orient* 46, no. 3 (2003): 320–354; Aida A. Essaid, *Zionism and Land Tenure in Mandate Palestine* (New York: Routledge, 2014), 103–107.

6. Marcel Mauss, "Techniques of the Body," *Economy and Society* 2, no. 1 ([1936] 1973): 70–88; Pierre Bourdieu, *Outline of a Theory of Practice*, trans. R. Nice (Cambridge: Cambridge University Press, 1977).

7. For more on the tiyul, see Orit Ben-David, "Tiyul as an Act of Consecration of Space," in *Grasping Land: Space and Place in Contemporary Israeli Discourse and Experience*, ed. Eyal Ben-Ari and Yoram Bilu (Albany: SUNY Press, 1997); Maoz Azaryahu and Arnon Golan, "Zionist Homelandscapes (and Their Constitution) in Israeli

Geography," *Social and Cultural Geography* 5, no. 3 (2004): 497–513; Shaul Kelner, "Ethnographers and History," *American Jewish History* 98, no. 1 (2014): 17–22. On Palestinian hill walking as a way of knowing the land, see Raja Shehadeh, *Palestinian Walks: Notes on a Vanishing Landscape* (London: Profile, 2007); Hayden Lorimer, "Walking: New Forms and Spaces for the Study of Pedestrianism," in *Geographies of Mobilities: Practices, Spaces, Subjects*, ed. Tim Cresswell (Burlington, VT: Ashgate Publishing, 2011), 25; David Pinder, "Errant Paths: The Poetics and Politics of Walking," *Environment and Planning D: Society and Space* 29, no. 4 (2011): 672–692.

8. Dov Beker, *In the Meadows* (Association of Shepherds in Israel, 1972), 5.

9. Efraim Eliash, "We Moved to Hamara," in *The Hebrew Shepherd*, ed. David Zamir and Matityahu Shelem (Merhavia: Hebrew Shepherds Association, 1957), 47.

10. Dr. Siegfried Hirsch, "Sheep and Goats in Palestine," *Bulletin of the Palestine Economic Society* 6, no. 2 (1933): 12, 32. As opposed to the Western conception (which was common among Jewish settlers as well), shepherding was not Bedouins' main occupation from the beginning of time. Anthropologist Dan Rabinowitz discusses how shepherding among Bedouins in Sinai occupied a secondary role throughout the nineteenth century, but was adopted again when other sources of income were no longer available. Dan Rabinowitz, "Themes in the Economy of the Bedouin of South Sinai in the Nineteenth and Twentieth Centuries," *International Journal of Middle East Studies* 17, no. 2 (1985): 211–228. Seth Frantzman and Ruth Kark similarly describe changes in Bedouin practices related to shifts in patterns of settlements in the twentieth century. Seth J. Franzman and Ruth Kark, "Bedouin Settlement in Late Ottoman and British Mandatory Palestine: Influence on the Cultural and Environmental Landscape, 1870–1948," *New Middle Eastern Studies* 1 (2011): 1–23.

11. On the role of sheep in the European expansion of the Americas, see Elinor G. K. Melville, *A Plague of Sheep: Environmental Consequences of the Conquest of Mexico* (New York: Cambridge University Press, 2005). On the role of sheep breeding for British colonialism, see Sarah Franklin, *Dolly Mixtures*; Rebecca J. H. Woods, *The Herds Shot Round the World: Native Breeds and the British Empire, 1800–1900* (Chapel Hill: University of North Carolina Press, 2017).

12. David Zamir, "On the Design of the Figure of the Hebrew Shepherd," lecture, July 28, 1952, 2, Hashomer Hatzair Archives, Yad-Ya'ari, Givat Haviva [hereafter HHA], 125.8.

13. Different Europeans, both Christians and Jews, lived with Arabs, whether peasants and pastoralists, in order to study them during the first decades of the twentieth century. Four examples of researchers who then published extensively are Finnish anthropologist Hilma Granqvist (1890–1972) who lived with peasants in Artas (see chapter 1), Jewish settler Tuvia Ashkenazi (1904–1970) who lived with different tribes of Bedouins, shepherd and archaeologist Pessah Bar-Adon (1907–1975) who lived with the Bedouin initially for the sake of conducting anthropological studies (see chapter 3, p. 82 [where the story of Bar-Adon and his image are discussed]),

and Orientalist and diplomat Eliyahu Epstein (Eilat) (1903–1990). All three Jewish settlers—Ashkenazi, Bar-Adon, and Epstein—were also involved in intelligence work for different Jewish organizations. See Hillel Cohen, *Army of Shadows: Palestinian Collaboration with Zionism, 1917–1948*, trans. Haim Watzman (Berkeley: University of California Press, 2008), 76, 115; Tuvia Ashkenazy, *The Bedouins in the Land of Israel* (New York: Ariel Books, 1957); Pessah Bar-Adon, *In Desert Tents: From the Notes of a Hebrew Shepherd among Bedouin Tribes* (Jerusalem: Kiryat-Sefer, 1981); Eliyahu Eilat (Esptein), *The Bedouins: Their Lives and Customs* (Tel Aviv: A. Y. Stibel, 1933). For a discussion on "biblical parallelism" among Palestinian intellectuals such as Tawfiq Canaan and Stephan Hanna Stephan, see Salim Tamari, *Mountain against the Sea: Essays on Palestinian Society and Culture* (Berkeley: University of California Press, 2009), 96–98.

14. The representation of the "New Jew" as well as the figure of the Hebrew shepherd was always gendered male. For further discussion, see chapter 4.

15. On the choice of the Bedouin as a model of the heroic native, see Yael Zerubavel, "Memory, the Rebirth of the Native, and the 'Hebrew Bedouin' Identity," *Social Research* 75, no. 1 (2008): 315–352.

16. Yaakov Goldstein, *The Shepherds Fraternity* (Tel Aviv: Ministry of Defense Publications, 1993), 13; Michal Sadan, *The Hebrew Shepherd: Transformation of Image and Symbol from the Hebrew Enlightenment Literature to the New Hebrew Culture in Israel* (Jerusalem: Yad Ben-Zvi, 2011), 38.

17. The few goats had a special role in Bedouin herds, as expediters of the sheep.

18. Quoted in David Zamir and Matityahu Shelem, eds., *The Hebrew Shepherd* (Merhavia: Hebrew Shepherds Association, 1957), 99.

19. M. Livne, "Lecture for the Course of Shepherds Training, 16–20 October 1951," HHA, 125.8.2.

20. Quoted in Zamir and Shelem, *The Hebrew Shepherd*, 100.

21. David Zamir, quoted in in Zamir and Shelem, *The Hebrew Shepherd*, 70.

22. Moshe, quoted in Zamir and Shelem, *The Hebrew Shepherd*, 43.

23. Dov Blumberg, "Kfar Gila'di, Early January 1922," Dov Blumberg's diary, Dov Blumberg Collection, Lavon Institute for the Study of the Labor Movement, Tel Aviv, IV-104-26-4. I thank Matan Boord for this document.

24. There are similarities between the Hebrew shepherds and "Canaanism," a group of settlers that aspired to revive ancient Canaanite life in Palestine in the 1940s. But as opposed to the shepherd community, which attempted to revive Hebrew life by way of practice and exploration of the land, Canaanism was an urban and intellectual phenomenon, focusing mainly on poetry. As Yaacov Shavit demonstrates, the members of that movement did not seek intimate knowledge and connection with the environment. For them, the landscape was not a concrete thing but rather

merely a metaphor. Yaacov Shavit, *From Hebrew to Canaanite: Aspects in the History, Ideology and Utopia of the "Hebrew Renaissance"—From Radical Zionism to Anti-Zionism* (Jerusalem: Domino Press, 1984), 42–43.

25. For a description of these failing attempts, see Moshe, quoted in Zamir and Shelem, *The Hebrew Shepherd*, 43. One famous example of failure was the shepherds group at Sheikh Abrek hills in the second half of the 1920s. See HHA, IV-235-1-2536, 42–92. Another was the unrealized cooperative shepherds group at Poria. See HHA, IV-235-2-80.

26. For an analysis of the failing attempts of the first generation of Jewish shepherds and the shepherds' relation to the Hashomer movement, see Goldstein, *The Shepherds Fraternity*.

27. Quoted in Zamir and Shelem, *The Hebrew Shepherd*, 72.

28. In *The Hebrew Shepherd*, shepherds recount several cases when the newly trained shepherds did not succeed in keeping their beloved sheep healthy or even alive. One example was a herd that died only few days after arriving at the settlement at Kinneret. Zamir and Shelem, *The Hebrew Shepherd*, 43.

29. Moshe Jlodin, quoted in Zamir and Shelem, *The Hebrew Shepherd*, 45.

30. Quoted in Zamir and Shelem, *The Hebrew Shepherd*, 47.

31. In various letters and reports, the shepherds complain about the Jewish leadership's favoring of cow milk production over sheep milk and the work of the Cattle Breeders Association over that of the shepherds community. For further discussion of the centrality of cow milk to the settler economy, see chapter 4.

32. Zamir, "On the Design of the Figure of the Hebrew Shepherd."

33. Mordechai Livne, in Zamir and Shelem, *The Hebrew Shepherd*, 121.

34. Interview with 101-year-old former shepherd Lotek Etsion, Kibbutz Merhavia, October 18, 2011; interview with Yosefa Pecher, whose father was a shepherd in the 1940s, Kibbutz Mizra', February 9, 2012. Organized training with sheep in agricultural schools for youths began in 1934. Zamir and Shelem, *The Hebrew Shepherd*, 200.

35. For the centrality of the *bryndza* cheese as part of Jewish sheep management, see Hirsch, "Sheep and Goats in Palestine," 40.

36. Dov Beker, quoted in Zamir and Shelem, *The Hebrew Shepherd*, 166.

37. Shepherd Pima writes in 1950, "We got the local sheep and developed the traits for enhanced milk production. There were also other suggestions—the direction of meat and wool. It was not coincidental that we went in the direction of developing milk traits." Quoted in minutes of the twenty-first meeting of the Shepherds Association, November 12, 1950, Lavon Institute for Labor Research Archives [hereafter LILR], IV-280–1.

38. "Seasonal Instruction for the Raising of Sheep and Goats," Ministry of Agriculture, Animal Division, Sheep and Goat Department, January 16, 1949, HHA, 125.2.2.

39. Zamir and Shelem, *The Hebrew Shepherd*, 132. I found that in a report in English by the association, the name was translated into the Jewish Sheep Breeders, perhaps imitating the Dairy Cattle Breeders Association.

40. Zamir and Shelem, *The Hebrew Shepherd*, 210, 137.

41. For acknowledgments of the receipt of various publications (shepherds in China, the USSR, and New Zealand to Aaron Harari), see LILR, IV-280–77. On the failure of crossbreeding with merino sheep, see Dr. Henry P. Fox, M.R.C.V.S, "Sheep Husbandry and Goat Rearing in Jewish Settlements in Palestine," special investigation carried out between March 11, 1934 and May 17, 1934, 45, HHA, 125.3, 2.

42. Fox, "Sheep Husbandry and Goat Rearing in Jewish Settlements in Palestine," 49.

43. Zamir and Shelem, *The Hebrew Shepherd*, 142. See also figure 3.4.

44. Letter from Hebrew Shepherds Association to Mr. M. Beder at the Agricultural Center, Tel Aviv, September 24, 1942, HHA, 125.2.3.

45. Zamir, "On the Design of the Figure of the Hebrew Shepherd."

46. Beker (*In the Meadows*, 5) writes, "Our fathers were livestock people . . . and like the shepherd, also his sheep. The sheep that we have been improving in this land are an inheritance from the days of the fathers—[they are] the sheep of the fathers." It was important for Jewish shepherds to use the local breed, which they considered ancient, but its Arab name was problematic. How could Hebrew shepherds rear Arab sheep? The shepherds acknowledged the significance of the name of the breed to their practice and debated what other name might be most appropriate. Berl, a member of the association, told his colleagues in their meeting in 1950, "There was an objection to the name 'Awasi,' and we searched for a name that would fit the best quality of sheep of our environment. The committee decided to broaden the name to the entire breed—[and call it] 'the Hebrew breed.' Our fathers predated the Arabs who gave the name Awasi. Our fathers developed sheep rearing and improved the ancient stock. Thanks to them, we are entitled to call the sheep and the breed 'Hebrew.'" Not everyone agreed, however. Another member, Israel Ben-Shem, objected to the use of the name "Hebrew" since it did not capture the improvement of the breed done by Hebrew shepherds. Minutes of the twenty-first meeting of the Shepherds Association.

47. Pima, quoted in minutes of the twenty-first meeting of the Shepherds Association.

48. On the artistic representation of the shepherd in Jewish culture, see Michal Sadan, *The Hebrew Shepherd: Transformation of Image and Symbol from the Hebrew Enlightenment Literature to the New Hebrew Culture in Israel* (Jerusalem: Yad Ben-Zvi, 2011). Three famous painters who were also shepherds are David Alef (Alkind) (1908–?) of Kibbutz Beit-Alfa, later Ramat-Yohanan, and finally Beit-Hashita, Aharon Harari (1908–1984) of Kibbutz Merhavia, and Leo Roth (1914–2002) of Kibbutz

Afikim. In addition to being a shepherd himself, Roth invited Siegfried Shalom Sebba (1897–1975), another renowned artist, to stay with him and witness the practice of shepherding. This long visit is known to have influenced Sebba deeply and resulted in one of the most important Jewish/Israeli paintings, *The Shearing (Hagez)*, in 1947.

49. See, for example, the paintings of Sebba as well as Ze'ev Raban from the Bezalel School of Art, Nahum Gutman, and Menashe Kadishman.

50. Leila Abu-Lughod, *Veiled Sentiments: Honor and Poetry in a Bedouin Society* (Berkeley: University of California Press, 1986). The aboriginal "songlines" is a practice that combines singing with a knowledge of the land. See Bruce Chatwin, *The Songlines* (New York: Penguin Books, 1987).

51. Matityahu Shelem, "Know Dear Shepherd," Zemereshet, http://www.zemereshet .co.il/song.asp?id=654.

52. Matityahu Shelem, "The Sheep Have Spread," Zemereshet, http://www.zeme reshet.co.il/song.asp?id=1688.

53. Ruth Eshel, "Dancing on the Sidewalks of Ein-Hashofet," *Dance Now* 14, no. 1 (2000): 44–48.

54. Mordechai Amitai, quoted in Zamir and Shelem, *The Hebrew Shepherd*, 222.

55. "The Shearing Holiday and Its Origins," cited in Zamir and Shelem, *The Hebrew Shepherd*, 223.

56. "Report on the Activities of the Hebrew Shepherds Association for Its 25th Anniversary," 1954, HHA, 25.2.2, 3.

57. M. Livne, quoted in Zamir and Shelem, *The Hebrew Shepherd*, 171.

58. See letter from Moshe Sheler, Tel Aviv, to Aharon Harari, Merhavia, May 30, 1948, KMA, Aharon Harari Collection, 2.7; letter from the Hebrew Shepherds Association to the shepherds of Merhavia, July 25, 1948, KMA, Aharon Harari Collection, 2.6. Both of these letters were dealing with booty sheep held by the Israeli military and offered to Jewish settlements. For complaints regarding the confiscation of sheep, goats, and cows by the Israel Defense Forces as well as formal estimations, see Israel State Archives [hereafter ISA], Gimel-14, 309.

59. The purchase of sheep across borders was coordinated between the Hebrew Shepherds Association, prime minister's office, Ministry of Agriculture, and police. See ISA, Gimel-5, 17117.

60. State support for the purchasing of sheep from across borders happened concurrently to the efforts to reduce the number of herds owned by Palestinians. Palestinian attempts to purchase sheep from across the borders were defined as "smuggling" and entailed punishment. See ISA, Gimel-10, 2865.

61. See Central Zionist Archives, S, 15, 40983.

62. "Summary of the Activities of the Passing Year," *Hanoked* 1 (1953), HHA, Hanoked Journal.

63. "Report on the Activities of the Hebrew Shepherds Association for Its 25th Anniversary," 3.

64. David Eydlin, "Sheep Management," January 1991, 2, KMA, 5.5: Sheep. "Not all agricultural sectors can solve their problems with machines," said one shepherd in 1950. Quoted in minutes of the twenty-first meeting of the Shepherds Association.

65. In a meeting with the Hebrew Shepherds Association in 1954, the manager of Tnuva said that "by no means will Tnuva agree to the mixing of sheep with cows milk!" Quoted in minutes of the meeting, March 29, 1954, HHA, 125.2.3. See also Tnuva to Kibbutz Merhavia, March 10, 1941, KMA, 5.5.1 (sheep 1930–1959), 2; Zamir and Shelem, *The Hebrew Shepherd*, 158.

66. Ronny Gitter, "'Wolves' Are Needed for the Herds of Sheep," 3, HHA, 125.12.1.

67. Eydlin, "Sheep Management."

68. "Summary of the Activities of the Passing Year."

69. Pessah Bar-Adon, "A Sheep," *Mozna'im: Journal for Literature, Criticism, and Art* 3 (November 3, 1933): 2–4; Aziz Afandi [Pessah Bar-Adon], *Among the Herds of Sheep (From the Stories of a Shepherd)* (Tel Aviv: Eli'asaf Publishers, 1942).

70. See, for example, the description of a communal vote by Rivka Gorfien in Zamir and Shelem, *The Hebrew Shepherd*, 289; minutes of the twenty-first meeting of the Shepherds Association.

71. Zamir, "On the Design of the Figure of the Hebrew Shepherd."

72. "David Zamir," biographical notes.

73. Quoted in Zamir and Shelem, *The Hebrew Shepherd*, 164.

74. H. Horowitz, quoted in Zamir and Shelem, *The Hebrew Shepherd*, 213.

75. As part of their efforts to make sheep rearing profitable, agricultural experts in Israel began to develop a dual-purpose sheep (raised for both milk and meat) in the mid-1950s. The Assaf, a crossbreed of the local Awasi and East Friesian sheep, became the dominant breed on Jewish farms in the 1970s. In recent years, most Jewish farmers raise sheep for their meat (approximately 85 percent). On kibbutzim today, there are only 7 sheep-raising farms as opposed to about 120 in the 1950s. This data is based on my correspondence with Dorit Kababia, manager of the Sheep and Goats Section, Ministry of Agriculture, Beit Dagan, July 13, 2014, and my correspondence with Yosef Carasso, manager of the Sheep and Goats Section during 1970–1999, August 17, 2014.

76. Another indication of this impossibility lies in the Hebrew language. As opposed to the Hebrew word for a cow farmer, *Raftan* (literally, a man who works in the cowshed), or beekeeper, *Dvorai* or even *Kavran* (literally, a man who works with hives), a

word doesn't exist for a sheep farmer (which might be *Diran*, a person who works in the pen). In contrast, there are two Hebrew words for shepherd.

77. Tal Elmaliach, *The Kibbutz Industry, 1923–2007: Discussing Questions of Economy and Society* (Givat Haviva: Yad Ya'ari Publishers, 2009).

78. On the forced resettlement of Bedouin in Israel, see, for example, Ghazi Falah, "How Israel Controls the Bedouin in Israel," *Journal of Palestine Studies* 14, no. 2 (1985): 35–51; Ahmad Amara, Ismael Abu-Saad, and Oren Yiftachel, eds., *Indigenous (In)Justice: Human Rights Law and Bedouin Arabs in the Naqab/Negev* (Cambridge, MA: Harvard University Press, 2012).

79. Bedouin shepherding did not go totally extinct under Israeli rule. Anthropologist Aref Abu-Rabia demonstrates today how against all odds, and in contrast to the state efforts of sedentarization, a significant number of Bedouins continue to hold sheep and goats as well as practice shepherding. Yet sheep rearing does not occupy the same central role as it had before, and families no longer rely on sheep rearing as a main occupation and source of income. Moreover, most owners of sheep do not usually herd them themselves but instead hire others to be shepherds. Aref Abu-Rabia, *The Negev Bedouin and Livestock Rearing: Social, Economic, and Political Aspects* (Oxford: Berg Publishers, 1994).

80. In a 2011 interview with Lotek Etsion, he noted the various careers he had held since the closure of the pen at Merhavia, but also said that there were many shepherds that later became "important [not necessarily medical] doctors."

CHAPTER 4

1. Itzhak Elazari-Volcani, *The Dairy Industry as the Basis for Colonisation in Palestine* (Tel Aviv: Palestine Economic Society, 1928).

2. "And These Are the Happenings of Stavit," in *Stavit: For the Summary of a Yield of 100,00 Kg of Milk—in the Conference for the Crowning of Stavit* (Tel Aviv: Israeli Breeders Association, October 12, 1950).

3. Uriel Levi, *The History of Dairy Bovine in Israel* (Tel Aviv: Cattle Breeders Association in Israel, 1983), 200–202.

4. "'Senior Home' for Stavit," *Davar*, July 2, 1951, 4. For this cow's full biography, see Tamar Novick, "All about Stavit: A Beastly Biography," special issue, *Teoria Vebikoret* 51 (2019): 15–40.

5. "And These Are the Happenings of Stavit."

6. Interview with Dorit Kababia, manager of the Sheep and Goats Section, Ministry of Agriculture, Beit Dagan (September 15, 2013). For a recent incarnation of this claim about milk yield, see Ralph Ginsberg, "How Did the Israeli Holstein Cow Become a World Leader in Milk Yields?," Israel Dairy Board, https://www.israeldairy .com/israeli-holstein-cow-become-world-leader-milk-yields-3/; Israel Moshkovitz,

"World Record for the Israeli Cow: 11,400 Liter per Year," *Yedi'ot Achronot, Economy Supplement*, May 29, 2014, 10. I thank Noga Rosenfarb for this article.

7. Levi, *The History of Dairy Bovine in Israel*, 107–129.

8. Existing work deals with the role of dairy in the design of an economic model for Jewish agricultural settlements in the late Ottoman and British rules in Palestine. See Yaacov Shavit and Dan Gil'adi, "The Cowshed and the Agricultural Economy in Eretz Israel: The Place and Role of the Dairy Farming in the Jewish Settlement Program in Eretz Israel during the Mandate Period," *Catedra* 18 (1981): 178–193; Ayala Plezental, "'Milky Way': The Dairy Industry in Eretz Israel in the 1930s as a Mirror for German-Jewish Relations," in *Germany and Eretz Israel: Cultural Meeting Point*, ed. Moshe Zimmermann (Jerusalem: Hebrew University Magnes Press, 2003), 133–142. For studies on the development of the mixed-farming model in Palestine, see Derek Penslar, *Zionism and Technocracy: The Engineering of Jewish Settlement in Palestine, 1870–1918* (Bloomington: Indiana University Press, 1991); Hezi Amiur, *Mixed Farm and Smallholding in Zionist Thought* (Jerusalem: Zalman Shazar Center, 2016).

9. Melanie DuPuis, *Nature's Perfect Food: How Milk Became America's Drink* (New York: NYU Press, 2002); Kirk Kardeshian, *Milk Money: Cash, Cows, and the Death of the American Dairy Farm* (Lebanon: University of New Hampshire Press, 2012); Kendra Smith-Howard, *Pure and Modern Milk: An Environmental History since 1900* (Oxford: Oxford University Press, 2014); Deborah Valenze, *Milk: A Local and Global History* (New Haven, CT: Yale University Press, 2011); Sandra Aguilar-Rodríguez, "Nutrition and Modernity: Milk Consumption in 1940s and 1950s Mexico," *Radical History Review* 110 (2011): 36–58. Other works include Micah Aaron Rueber, "Making Milking Modern: Agriculture Science and the American Dairy, 1890–1940" (PhD diss., Mississippi State University, 2010); Heather Paxson, *The Life of Cheese: Crafting Food and Value in America* (Berkeley: University of California Press, 2013).

10. For a look at how the management of pedigree cows in Britain reflects changing social norms in nineteenth-century Britain, see Hariet Ritvo, *The Animal Estate: The English and Other Creatures in the Victorian Age* (Cambridge, MA: Harvard University Press, 1987). For an examination of the role of cows in the European colonization of New England in the seventeenth century and how it came to be that colonial cows went feral, see Virginia D. Anderson, *Creatures of Empire: How Domestic Animals Transformed Early America* (Oxford: Oxford University Press, 2004). See also Nicholas Russell, *Like Engend'ring Like: Heredity and Animal Breeding in Early Modern England* (Cambridge: Cambridge University Press, 1986); Hariet Ritvo, *Noble Cows and Hybrid Zebras: Essays on Animals and History* (Charlottesville: University of Virginia Press, 2010); Rebecca J. H. Woods, *The Herds Shot Round the World: Native Breeds and the British Empire, 1800–1900* (Chapel Hill: University of North Carolina Press, 2017).

11. See Emily Pawley, "The Point of Perfection: Cattle Portraiture, Bloodlines, and the Meaning of Breeding," *Journal of the Early Republic* 36, no. 1 (2016): 37–72; Woods, *The Herds Shot Round the World*; Gabriel N. Rosenberg, "No Scrubs: Livestock

Breeding, Eugenics, and the State in Early Twentieth-Century United States," *Journal of American History* 107, no. 2 (2020): 362–387. On the breeding of dairy cattle, see Barbara Orland, "Turbo-Cows: Producing a Competitive Animal in the Nineteenth and Early Twentieth Centuries," in *Industrializing Organisms: Introducing Evolutionary History*, ed. Susan R. Schrepfer and Philip Scranton (New York: Routledge, 2003), 167–189; Bert Theunissen, "Breeding for Nobility or for Production? Cultures of Dairy Cattle Breeding in the Netherlands, 1945–1995," *Isis* 103, no. 2 (2012): 278–309. The connection between milk and the reproduction of cows and women was emphasized by ecofeminist Greta Gaard, who called for the establishment of "feminist postcolonial milk studies." Greta Gaard, "Toward a Feminist Postcolonial Milk Studies," *American Quarterly* 65, no. 3 (2013): 595–618.

12. Historian Derek Penslar shows how such figures were neither only policy makers nor experts or agriculturalists, but some combination of the three. Derek Penslar, *Israel in History: The Jewish State in Comparative Perspective* (New York: Routledge, 2007), 154–159.

13. Scholars' discussions of this phenomenon are always human centered and composed of some combination of three explanations: the Jewish commandment to "be fertile and increase" (*pru urvru*), the regional demographic fear, and aspirations to rehabilitate the Jewish people after the Holocaust. US researchers studying this phenomenon seem to have paid most attention to the Jewish aspect, dealing, for example, with how technomedical interventions intersect with religious ideas that they consider to be deeply ingrained in the Israeli state, and how medical and religious institutions interact. See Susan Martha Kahn, *Reproducing Jews: A Cultural Account of Assisted Conception in Israel* (Durham, NC: Duke University Press, 2000); Elly Teman, *Birthing the Mother: The Surrogate Body and the Pregnant Self* (Berkeley: University of California Press, 2010). Israeli researchers seem to pay more attention to state support (pronatalist policies such as the initiation of the Motherhood Prize, which gratifies every Israeli mother for raising ten children or more) and the familial sensibility in Israeli society that is attributed to a combination of Jewish ideals and "traditional" ways of living. See, for example, Sylvia Fogel-Bijawi, "Families in Israel: Postmodernity, Feminism and the State," *Journal of Israeli History* 21, no. 2 (2002): 38–62; Daphna Birenboim-Carmeli and Yoram Carmeli, eds., *Kin, Gene, Community: Reproductive Technologies among Jewish Israelis* (New York: Berghahn Books, 2010); Yael Hashiloni-Dolev, *The Fertility Revolution* (Ben-Shemen: Modan, 2013). Some work discusses the problems of infertility and fertility treatments among Palestinian women. See, for example, Rhoda Ann Kanaaneh, *Birthing the Nation: Strategies of Palestinian Women in Israel* (Berkeley: University of California Press, 2002); Livia Wick, "Making Lives under Closure: Birth and Medicine in Palestine's Waiting Zones" (PhD diss., Massachusetts Institute of Technology, 2006); Himmat Zu'bi, "Palestinian Fertility in the Israeli Sphere: Palestinian Women in Israel Undergoing IVF Treatments," in *Bioethics and Biopolitics in Israel: Socio-Legal, Political, and Empirical Analysis*, ed. Hagai Boas, Yael Hashiloni-Dolev, Nadav Davidovitch, Dani Filk, and Shai J. Lavi (Cambridge: Cambridge University Press, 2018), 160–180.

14. Elimelech Zagorodsky, "Milk and Its Outcomes," *Hameshek Hahaklai* 3, no. 3 (1914), 65. The same text also appears in Elimelech Zagorodsky, *The Book of Milk, Part I: Milk and Its Outcomes* (Jaffa: Hachak'lai Publishers, 1914), 1. I thank Efrat Gilad for this book.

15. For an exploration by an agricultural engineer demonstrating the prevalence of the dual use of cattle (as both beasts of burden and milkers) among Palestinian Arabs as well as for information on the desired qualities of a milking cow, see, for example, Wasfi Beck Zakkariyah, "Cattle Raising," *Al-Iqtisadat al-'Arabiyya*, December 2, 1936, 8–9.

16. For descriptions of settlers indicating the prevalence of fresh goat milk (and goats) and how hard it was to find cow milk in the cities in early twentieth-century Palestine, see Shlomo Dori, *News from the Past: Chapters in the History of Dairy Cattle Farming in Israel, Part 2* (Caesarea: Cattle Breeders Association in Israel, 1996), 79, 105, 107; Shlomo Dori, *News from the Past: Chapters in the History of Dairy Cattle Farming in Israel, Part 3* (Caesarea: Cattle Breeders Association in Israel, 2002), 75. With the growing European settlement in Palestine, goats became a major threat to the project of afforestation, and the British and then Israeli governments issued a series of laws limiting their movement and eating habits (see chapter 2).

17. "Deep uddered kine" is a biblical phrase that refers to the breasts of cows.

18. Ada Goodrich-Fereer, *Arabs in Tent and Town: An Intimate Account of the Family Life of the Arabs of Syria, Their Manner of Living in Desert and Town, Their Hospitality, Customs and Mental Attitude, with a Description of the Animals Birds, Flowers and Plants of Their Country* (New York: G. P. Putnam's Sons, 1924), 142. Moshe Palmon of the Breeders Association remembers that "we did not have cattle in this country. None." Moshe Palmon, interview by Nir Mann, *Efraim Smaragd*, Tel-Aviv, May 27, 2001. This understanding of the land as empty of cows fits the general colonial perspective of terra nullius and hence the title of this section. In most other cases, however, terra nullius did not fit the way that European settlers viewed Palestine. For several reasons—among which is the need to find explanations for the failure of the cru-sades as well as the choice to see Palestinians as the link to biblical Hebrews (see chapter 3)—Europeans in Palestine did not tend to see the land as empty.

19. For a discussion in Arabic of the importance of milk to all people, see Ibn Al-Sheikh, "Milk—Important Information for Farmers and Non-Farmers," *Al-Liwaa*, February 18, 1936, 7.

20. It is important to note that various accounts talk about the importation of cows for slaughter during the British rule. While the raising of cows for meat has been enormously important in global imperialism, and the British Empire in par-ticular, settlers (especially Jewish settlers) put very little emphasis on cow meat. This might be partially explained by the limited pasture area in Palestine. The major-ity of beef and mutton were imported and not raised in Palestine. The Beef Cattle Breeders Association, furthermore, was established as late as 1956, compared to the

establishment of the Dairy Cattle Breeders Association in 1926. On meat consumption among Jewish settlers in British Palestine see Efrat Gilad, "Meat in the Heat: A History of Tel Aviv under the British Mandate for Palestine (1920s–1940s)," PhD diss., University of Geneva, 2021.

21. See Anderson, *Creatures of Empire*.

22. Paul Sauer, *The Holy Land Called: The Story of the the Temple Society* (Stuttgart: Konrad Theiss Verlag GmbH, 1985), 11, 18. For more information on the Templar settlement in Palestine, see Alex Carmel, *The German Settlement in the Land of Israel in the Late Ottoman Period: National, Local, and International Problems* (Haifa: Haifa University Press, 1970).

23. For German research projects in Palestine, see Haim Goren, *Go Research the Land: German Studies of the Land of Israel in the Nineteenth Century* (Jerusalem: Yad Ben-Zvi, 1999).

24. Naftali Talmon, "The Agricultural Farm in the Templar Settlements and Its Contribution for the Development of Agriculture in Eretz Israel," *Catedra* 78 (1995): 69.

25. There are a few indications that the crossings of European and local breeds existed in other locations in late nineteenth-century Palestine, such as in several monasteries in the Jerusalem area and Mikve-Israel, the Jewish agricultural school founded 1870 that belonged to the Alliance Israélite Universelle and was largely funded by French philanthropist Edmond James de Rothschild. Shlomo Dori, *News from the Past: Chapters in the History of Dairy Cattle Farming in Israel, Part 1* (Caesarea: Cattle Breeders Association in Israel, 1992), 18, 98.

26. Goren, *Go Research the Land*, 234–235; Haim Goren, *True Catholics and Good Germans: The Catholic Germans and the Land of Israel, 1838–1910* (Jerusalem: Hebrew University Magnes Press, 2004), 55–57; Talmon, "The Agricultural Farm in the Templar Settlements," 65–81.

27. Various contemporary and historical works discuss the Jewish study and adoption of German Templar practices along with the ongoing cooperation between these communities. For the connections between Templar dairy farming and Jewish settlement, see Naftali Talmon, "Fritz Keller—Of the Pioneers of Modern Agriculture in Eretz Israel," in *The Landscape of His Homeland: Studies in the Geography and History of Eretz Israel (Dedicated to Yehosuah Ben-Arieh)*, ed. Yossi Ben-Artzi, Israel Bartal, and Elhana Riner (Jerusalem: Hebrew University Magnes Press, 1999), 333–351. For a look at the cooperation in the dairy industry between Jewish settlers and German Templars throughout the 1930s, in spite of restrictions issued by the Jewish leadership following the rise of the Nazi rule in Germany and the growing propaganda for purchasing local Jewish-made products as part of the Product of the Land campaign, see Ayala Plezental, "'Milky Way': The Dairy Industry in Eretz Israel in the 1930s as a Mirror for German-Jewish Relations," in *Germany and Eretz Israel: Cultural Meeting Point*, ed. Moshe Zimmermann (Jerusalem: Hebrew University Magnes Press, 2003), 133–142. Furthermore, in 1911, Yitzhak Wilkansky (later Elazari-Volcani) wrote that

"Wilhelma [a Templar settlement] should be a model for us." Yitzhak Wilkansky to the Land of Israel Office, World Zionist Organization, May 18, 1911, Archives of the Dairy Cattle Breeders Association, Kibbutz Yif'at [hereafter DCBA], 7, 4, 1, 4.

28. Levi, *The History of Dairy Bovine in Israel*, 76. Estimations of the numbers of cows vary, as cows were analyzed differently—differentiated at times between dairy and other cattle, between cows and bulls, and between cattle owned by Jewish settlers or others. Shmuel Avitzur used the Ottoman and British censuses and estimated that there were 58,000 heads cattle in 1914, and then 82,480 in 1922. According to Salo Jones and based on the earliest British census, there were overall 80,000 heads of cattle in 1921. According to E. Ray Casto, there were 160,000 in 1932. J. M. Smith and S. J. Gilbert recorded 160,000 native cattle and 14,000 dairy cattle in 1934. For the Jewish population, Avitzur documented 1,300 in 1900, 5,808 in 1922, 11,521 in 1927, 17,994 in 1937, and 30,836 in 1944. Moshe Yoel Pinthus noted that there were 6,300 in 1922, 18,000 in 1936, and 34,000 in 1948. Haim Halperin reported that there were 36,043 in 1947 and 69,781 in 1953. Shmuel Avitzur, *Changes in Eretz-Israel Agriculture, 1875–1975* (Tel Aviv: Milo Ltd. and Avshalom Institute, 1977), 209; Salo Jonas, "Cattle Raising in Palestine," *Agricultural History* 26, no. 3 (1952): 94; E. Ray Casto, "Economic Geography of Palestine," *Economic Geography* 13, no. 3 (1937): 250; J. M. Smith and S. J. Gilbert, "Cattle Disease Occurring in Palestine and Segregation and Prevention Control of Bovine Contagious Abortion," *Journal of Comparative Pathology and Therapeutics* 47 (1934): 94–106; Moshe Yoel Pinthus, *History of Agricultural Research in Israel in the Pre-Statehood Era (1920–1949)* (Haifa: Itay Bahur, 2011), 329; Haim Halperin, *Changes in Jewish Agriculture* (Tel Aviv: Ayanot, 1954), 56.

29. The Jewish Veterinary Services and Livestock Insurance (1920), Jewish Cattle Breeders Association (1926), and Central Cooperation for Agricultural Production in Palestine, or Tnuva for short (1926) were all established in the 1920s and 1930s, focusing their work on dairy cattle and cow milk. See Dori, *News from the Past, Part 1*, 54; Dori, *News from the Past, Part 2*, 38; Dori, *News from the Past, Part 3*, 5.

30. See "Food and Agricultural Commodity Production," Food and Agriculture Organization of the United Nations, http://faostat.fao.org/site/339/default.aspx.

31. Levi, *The History of Dairy Bovine in Israel*, 107–120.

32. According to Dori (*News from the Past, Part 2*, 59), a leading figure in the dairy farmer community, Jewish settlers tried and ultimately failed to use milking machines in 1935, as the machines caused too many problems, and "reduced the intimacy in the relationship between the cow and the milking person." The use of machines for milking, however, was picked up again in the mid-1940s and became prevalent in 1947.

33. People in Palestine were used to drinking goat milk. The dramatic growth of bovine milk production by both Templar and Jewish settlers resulted in competition over the limited milk markets in the cities. "It was hard for us to get used to cow milk, which only became a staple in the 1920s," remembered Dr. A. Shoshani.

Quoted in Dori, *News from the Past, Part 1*, 17–18. Similarly, the people of Kibbutz Degania recounted having problems selling their cow milk in the market of Tiberias. On his arrival to the city, Zvi Tamari of Degania "did not manage to find anyone that would buy the milk, because the women of Tiberias were used to go[ing] out to the street and buy[ing] milk from an Arab, who would milk his goat in their presence." Quoted in Dori, *News from the Past, Part 1*, 22. The dairy farmers of Kibbutz Mizra' complained about the same problem, describing the lack of a market for their milk due to "the competition with the Arab milk." Yosio Shemesh, "About the Cowshed," Kibbutz Mizra' Archives [hereafter MA], Journal Collection, *In Mizra' 1* (1937): 8.

34. Cow milk was the sixth-biggest agricultural commodity in Syria, the ninth in Egypt, and the nineteenth in Jordan in 1961. In 2007, cows in Israel produced 1.185 million liters of milk, with an estimated profit of US$3.13 million, making this commodity almost two times greater than that of potatoes, the runner-up. Such yield also made the dairy industry almost three times more profitable than the runner-up tomato industry. See "Food and Agricultural Commodity Production."

35. Over the years, the professionalization of cattle husbandry and the efforts to increase milk production have resulted in a wealth of research regarding the relation between milk production and the environment of Palestine/Israel. See, for example, Yeshayahu Folman, Amiel Berman, Zeev Herz, Moshe Kaim, Miriam Rosenberg, Meir Mamen, and Sali Gordin, "Milk Yield and Fertility of High-Yielding Dairy Cows in a Sub-Tropical Climate during Summer and Winter," *Journal of Dairy Research* 46, no. 3 (1979): 411–425; Miriam Rosenberg, Yeshayahu Folman, Zeev Herz, Israel Flamenbaum, Amiel Berman, and Moshe Kaim, "Effect of Climatic Conditions on Peripheral Concentrations of LH, Progesterone and Oestradiol-17b in High Milk-Yielding Cows," *Journal of Reproduction and Fertility* 66, no. 1 (1982): 139–146; D. Wolfenson, William W. Thatcher, Lokenga Badinga, J. D. Savio, Rina Meidan, B. J. Lew, Ruth Braw-Tal, and Amiel Berman, "Effect of Heat Stress on Follicular Development during the Estrous Cycle in Lactating Dairy Cattle," *Biology of Reproduction* 52, no. 5 (1995): 1106–1113; Advanced Cow Cooling Systems Co. and Hachaklait Veterinary Services Ltd., "Management of Heat Stress to Improve Fertility in Dairy Cows in Israel," *Journal of Reproduction and Development* 56 (2010): 36–51.

36. Tnuva became an important player in the Product of the Land project already mentioned, and promotion and sale of bovine milk and dairy products in particular. In 1938, in conjunction with the Natan Strauss Health House in Jerusalem and bacteriologist Israel Kilgler, Tnuva started the Glass of Milk a Day project. The supply of a glass of milk to schoolchildren (and for a while to factory workers) was based on a British model. See Nahum Verlinsky, *Debating Production* (Tel Aviv: Tnuva Center, 1973), 35–37. For a critique of the governmental Glass of Milk a Day program in the Palestinian Arab media, see "Machinery, Glass of Milk a Day, and Agricultural Experiments," *Al-Difa'*, April 9, 1943, 2. On plans to enhance milk consumption among Jewish settlers in Palestine see Efrat Gilad, "'The Child Needs Milk and Milk Needs a Market': The Politics of Nutrition in the Interwar Yishuv," *Gastronomica:*

The Journal for Food Studies 21, no.1 (2021): 7–16. An article from *Filastin* supplies a glimpse into the regional significance of milk and idea of drinking milk daily. In preparation for the visit of the son of Saudi king Abdulaziz Al Saud to Egypt in 1945, it reported, King Farouk gave an order to prepare a herd of goats and camels to enable the guest to drink milk daily during his weeklong stay. "Al-Farouk Receives the Son of Al Saud in the City of Suez: A Herd of Goats and Camels for Egypt's Great Visitor," *Filastin*, December 16, 1945, 1.

Dairy products have become important to Israelis. Soft cheeses, such as cottage cheese and another cheese known as "white cheese," are considered basic food staples and are found in most homes yearlong. In recent years, Tnuva, now a private company, used the centrality of milk symbolism to Israeli society to launch a campaign that equates its dairy products with "home." For a long while, people exiting Israel's national airport would witness an enormous Tnuva's cottage cheese "Welcome Home" sign. The privatization of Tnuva and sale of its stocks to foreign companies in 2006 resulted in an increase in the prices of dairy products. This rise in the cost of dairy products caused much rage in Israel, sparking the Israeli summer 2011 "social revolution," a series of demonstrations and acts of resistance to the government and its economic policies. In their early days, these acts of resistance were commonly called the "Cottage Cheese Revolution."

37. One example of the equating of dairy with Zionist movements is the Arab phrase *Arab al-Shamenet*, which non-Israeli Palestinians commonly use derogatively to refer to Palestinians who are citizens of Israel and critique their agreement to live under Israeli rule. *Arab al-Shamenet* means "Cream Arabs" and refers to the common consumption of sour cream among them—a dairy product that is industrially produced and holds a Hebrew name. Non-Israeli Palestinians see the common consumption of Israeli sour cream as a symbol of support for the Zionist economy and agenda, and hint to their privileged position. On the "move to milk" in early twentieth-century Britain and its national significance, see Sara Wilmot, "From 'Public Service' to Artificial Insemination: Animal Breeding Science and Reproductive Research in Early Twentieth-Century Britain," *Studies in History and Philosophy of Biological and Biomedical Sciences* 38 (2007): 411–441. For the importance of milk for Mexican nationalism, see Aguilar-Rodríguez, "Nutrition and Modernity." For milk and Cuban nationalism, see Reinaldo Funes Monzote, "Ubra Blanca and the Politics of Milk in Socialist Cuba," *Oxford Research Encyclopedia for Latin American History* (Oxford: Oxford University Press, 2019).

38. One example is Kiryat Anavim (est. 1920), an early Jewish communal agricultural settlement. Its cowshed, the first building constructed on Kiryat Anavim, had a Star of David on the entrance keystone. In this sense, Kiryat Anavim constructed a Jewish cowshed.

39. As many scholars have demonstrated, socialist thought and Eastern European agrarian culture enormously affected Jewish settlement patterns and agricultural practices in Palestine. I seek instead to identify the ideas and practices that emerged

within the process of settlement. Others have done the same. In his study of the labor market in late Ottoman Palestine, for example, sociologist Gershon Shafir shows that particular conditions in Palestine, not imported ideologies, gave rise to collective forms of life among Jewish settlers. In *Land, Labor and the Origins of the Israeli-Palestinian Conflict, 1882–1914* (Cambridge: Cambridge University Press, 1989).

40. For an elaborate analysis of Elazari-Volcani's part in the development of the mixed-farming model in Palestine, see Ami'or, "The Roots and Design of the 'Mixed-Farming Economy.'" The existing scholarship analyzing Elazari-Volcani's work has centered on the mixed-farming model and design of the semicommunal settlement, but not on the dairy industry, milk, or cows.

41. Israel Reichert, "To Yzhak Elazari-Volcani on His 70th Birthday: The Architect of the Settlement and Labor Zionism," *Davar*, February 3, 1950. The phrase "mentor for the settlement of the people on their land" was written on his tomb, which is located in the cemetery of Nahalal, the first semicommunal settlement.

42. The choice of pseudonyms hints at Elazari-Volcani's politico-religious agenda. The use of Ben-Abuya, the Talmudic "other" known for his heresy, probably signified Elazari-Volcani's desertion of Jewish faith and Jewish life in Europe for the sake of living a secular life in Palestine. The presentation of this contrast is fitting with a Zionist agenda, which altogether considered itself a transformative secular movement. His support for the Zionist idea is also implied by his other pseudonym, "A. Zionist." I thank Dan Tamir for his help in deciphering these pseudonyms.

43. Israel Reichert, "Yitzhak Wilkansky—for His 50th Birthday," *Davar*, February 2, 1931, 2. Boaz Neumann discusses the element of desire in the lives of Jewish settlers in the early twentieth century and how these settlers were influenced by European pastoral ideals. He looks at the writings of A. D. Gordon, who Jewish settlers considered a mentor. Elazari-Volcani was known for rejecting Gordon's ideas. See Boaz Neumann, *Land and Desire in Early Zionism* (Tel Aviv: Am Oved Publishers, 2009).

44. Yitzhak Elazari-Volcani, *On the Road* (Jaffa: Hapo'el Hatza'ir, 1918).

45. Yitzhak Elazari-Volcani, *The Design of the Agriculture of the Land (of Israel)* (Rehovot Experimental Station: Jewish Agency, 1937), 3–5.

46. This is a phrase from the book of Amos 7.14, meaning the one who will make the change happen.

47. Elazari-Volcani, *The Design of the Agriculture of the Land (of Israel)*, 5.

48. Elazari-Volcani, *The Design of the Agriculture of the Land (of Israel)*, 5. The use of the number sixty thousand cows is peculiar, as it does not fit any other estimates of either the number of heads of cattle among the general population or on Jewish farms. Using various other estimates, I assume that by using this number he refers to the population of female cows (rather than all cattle) in Palestine as a whole.

49. Issac Elazari-Volcani, *The Fellah's Farm* (Tel Aviv: Jewish Agency for Palestine, Institute of Agriculture and Natural History, Agricultural Experiment Station, 1930).

For a detailed analysis and critique of this project, see Omar Loren Tesdell, "Shadow Spaces: Territory, Sovereignty, and the Question of Palestinian Cultivation" (PhD diss., University of Minnesota, 2013), 1–72.

50. Historian Derek Penslar argues that Elazari-Volcani's (then going by Wilkansly) writing is "steeped in an odd brew of anti-industrialism and technophilia." Derek J. Penslar, *Israel in History: The Jewish State in Comparative Perspective* (New York: Routledge, 2007), 157–158. I contend that a complicated relationship to technologies, and mechanized technologies in particular, is inherent to settler colonial projects. See Tamar Novick, "The Hand, the Hoe, and the Hose: Contested Technologies of Settler Colonialism, in *The Cambridge Handbook for the History of Technology, Vol. II*, ed. Dagmar Schäfer, Francesca Bray, Matteo Valleriani, Tiago Saravia, and Shadreck Chirikure (forthcoming).

51. Elazari-Volcani was not alone. Accounts demonstrate that other settlers were hesitant to use new technologies on their farms, as the use of machines was seen as a threat to the intimate familiarity with the land. The members of Kibbutz Mizra', for example, in a series of articles titled "On Beast and Tractor" debated the utility of mechanization in the late 1930s. One settler voiced his concern regarding the distance drawn between settlers and nature, and called for the continued use of draft animals. Another noted in contrast that agriculture had nothing to do with nature and was instead a bread factory; in this sense, this settler, whose approach became the dominant one, contended that there was no need for romanticization apart from the romanticization of the machine. "On Beast and Tractor," MA, Journal Collection, *In Mizra'* 1 (1937): 2–3; "More about the Questions of the Beast and the Tractor," MA, Journal Collection, *In Mizra'* 2 (1937): 11. The understanding of farming as factory work rather than work in proximity to nature became prevalent in various other contexts and is reminiscent of Deborah Fitzgerald's work on the industrialization of agriculture in the United States. Deborah Fitzgerald, *Every Farm a Factory: The Industrial Ideal in American Agriculture* (New Haven, CT: Yale University Press, 2003).

52. Yitzhak Elazari-Volcani, "On the State of Farming," *Hapoel Hatzair*, February 11, 8. Elazari-Volcani clearly refers here to the Jewish people, connecting to the common idea that the Jewish people will be redeemed through working the land. The broader scope of his work shows that he thought a change in the agricultural economy in Palestine would be beneficial to both Jews and Arabs who lived and worked the land. See, for example, Elazari-Volcani, *The Fellah's Farm*.

53. On the design of the Jewish agricultural economy in Palestine and Elazari-Volcani's specific role, see Saul Katz, "'The First Furrow': Ideology, Settlement, and Agriculture in Petah-Tikva in Its First Decade," *Catedra* 23 (1982): 124–157; Penslar, *Zionism and Technocracy*; Ilan S. Troen, *Imagining Zion: Dreams, Designs, and Realities in a Century of Jewish Settlement* (New Haven, CT: Yale University Press, 2003); Ami'or, "The Roots and Design of the 'Mixed-Farming Economy.'"

54. Y. Uri, "For the Memory of Yitzhak Volcani (a Year after His Death)," *Davar*, May 25, 1956, 5.

55. Reichert, "Yitzhak Wilkansky—for His 50th Birthday."

56. Quoted in Levi, *The History of Dairy Bovine in Israel*, 190.

57. Roza El-Eini, "The Implementation of British Agricultural Policy in Palestine in the 1930s," *Middle Eastern Studies* 32, no. 4 (1996): 211–250; Yossi Ben-Arzi, "Religious Ideology and Landscape Formations: The Case of the German Templars in Eretz-Israel," in *Ideology and the Landscape in Historical Perspective*, ed. Alan R. H. Baker and Gideon Biger (Cambridge: Cambridge University Press, 2006), 83–106.

58. El-Eini ("The Implementation of British Agricultural Policy") argues that British efforts targeted Arab peasantry in particular, yet Jewish settlers remained the main benefactors of these efforts for several reasons. While Elazari-Volcani thought, at least on paper, that his efforts were benefiting both Jewish Arab agricultural laborers in Palestine, Arab peasants were systematically losing dominance over the production of and profit from crops and all other industries. For exemplary Palestinian Arab complaints about discrimination as well as critiques of governmental policy in regard to milk production and supply, see "Machinery, Glass of Milk a Day, and Agricultural Experiments"; "About Cattle Tenders," *Al-Difa'*, March 4, 1945, 3; "Milk Farce in the Midst of Tragedy," *Filastin*, March 13, 1937, 1.

59. Quoted in Motti Zeira, *Man of Loves: The Story of Yehoshua Brandstetter* (Jerusalem: Yad Ben-Zvi, 2006), 25.

60. Zeira, *Man of Loves*, 44.

61. Yehoshua Brandstetter, "Memorandum regarding the Establishment of and Institute for the Breeding of Dairy Cattle in the Land of Israel," 1923, Central Zionist Archives, S15, 21032.

62. Zeira, *Man of Loves*, 78, 80–81.

63. Brandstetter made *Land of Promise* with his second wife, Margot Klausner, who was to become an influential filmmaker, and Juda Leman, who directed the film. F. S. N., "Land of Promise at the Astor," *New York Times*, November 21, 1935. Following the enthusiastic US reception of the film, the *New York Times* argued that "if further proof be needed of the motion picture camera's ability to record history more vividly than any printed page, it may be found there."

64. *Land of Promise*, Spielberg Jewish Film Archive, minute 18:17, http://www.youtube.com/watch?v=QDoD6W2z01s.

65. Nadia Abu El-Haj, *Facts on the Ground: Archaeological Practice and Territorial Self-Fashioning in Israeli Society* (Chicago: University of Chicago Press, 2001). On genetic research that tried to prove such connections, see Rafael Falk, *Zionism and the Biology of the Jews* (Tel Aviv: Resling, 2006); Nadia Abu El-Haj, *The Genealogical Science: The Search for Jewish Origins and the Politics of Epistemology* (Chicago: University of Chicago Press, 2012).

66. See, for example, the story of Miriam Baratz in the introduction. Furthermore, Elazari-Volcani (*The Fellah's Farm*) himself was writing about the importance of learning from the fellah's experiences of hundreds of years.

67. A wealth of research was devoted to the growth of ideas of the New Jew, "Muscular Jew," and "Sabra," and the roles the emerging ideal body played for the Zionist settlement project. See Oz Almog, *The Sabra: The Creation of the New Jew*, trans. Haim Watzman (Berkeley: University of California Press, 2000); Meira Weiss, *The Chosen Body: The Politics of the Body in Israeli Society* (Stanford, CA: Stanford University Press, 2002); Falk, *Zionism and the Biology of the Jews*; Todd Samuel Presner, *Muscular Judaism: The Jewish Body and the Politics of Regeneration* (London: Routledge, 2007); Sandra M. Sufian, *Healing the Land and the Nation: Malaria and the Zionist Project in Palestine, 1920–1947* (Chicago: University of Chicago Press, 2007); Yael Zerubavel, "Memory, the Rebirth of the Native, and the "Hebrew Bedouin" Identity," *Social Research* 75, no. 1 (2008): 315–352; Neumann, *Land and Desire in Early Zionism*.

68. Michael Adas, *Machines as the Measure of Man: Science, Technology, and the Ideologies of Western Dominance* (Ithaca, NY: Cornell University Press, 1990); Joseph Morgan Hodge, *Triumph of the Expert: Agrarian Doctrines of Development and the Legacies of British Colonialism* (Athens: Ohio University Press, 2007).

69. Elazari-Volcani's (then Wilkansky) 1911 survey of dairies and dairy cows among European settlers shows that breeding practices existed, but were far and few between. Yitzhak Wilkansky to the Land of Israel Office, World Zionist Organization, May 18, 1911, DCBA, 7, 4, 1, 4.

70. An exemplary article dealing with the raising of dairy cattle details the way Jewish settlers viewed the Arab cow: "Wretched is the son of the land . . . and seven time[s] worse is his cow. . . . [F]rom the second year of her life . . . she produces little milk to revive the fruit of her womb, who runs around her, as she carries the plow. The peasant does not care for milk or dairy. When the Hebrew settlement wanted to create a dairy, it had to start from scratch. Because even the cow was not satisfactory." Moshe Sat-Ky, "A Short Guide for Raising Dairy Cattle," *Hasade* 17 (1937): 441. There are few indications showing that not all Jewish settlers denounced the Arab cow. See, for example, Stern from Ein-Harod, who posited in 1925 that "we cannot disregard the amount of milk we got from the Arab cows . . . we should be interested in the Arab cow. I am certain that an Arab cow is more profitable than a purebred cow," in Stern, "Finding Out a Method," *Hasade* 5 (1925).

71. Settler Tehia Liberson, a known women's rights activist, described her early work with cows in Nahalal, the first semicommunal agricultural settlement, in the 1920s. Liberson was assigned to work with an Arab cow that was exceptionally stubborn and wild, and it took much patience and suffering until they managed to get along. After she protested her terms of sustenance, which she attributed to her being a woman, she finally managed to better her situation: she got a plot of land to farm, and the Arab cow was substituted for a Damascene, which produced much more milk. Dori, *News from the Past, Part 3*, 13.

72. Sir John Hope Simpson, *Palestine: Report on Immigration, Land Settlement and Development* (London: Secretary of State for the Colonies, 1930), 77; El-Eini, "The Implementation of British Agricultural Policy," 227.

73. Yafa Berlovitch, *Wounded Bird: Dora Bader—A Diary (1933–1937)* (Kibbutz Mizra': Hakibbutz Hame'uchad Publishers, 2011), 342.

74. Pinthus, *History of Agricultural Research in Israel*), 331.

75. For cattle population statistics, see note 28.

76. Pinthus, *History of Agricultural Research in Israel*, 332. The statistics of milk yields vary according to different accounts, yet they all point to dramatic growth, moving from six to eight hundred liters per year in the early 1920s, to three to four thousand liters annually in the late 1930s. Avitzur, *Changes in Eretz-Israel Agriculture*, 209, 211.

77. See Theunissen, "Breeding for Nobility or for Production?" The British dairy culture concentrated on milk production around the same period; for example, the British Milk Marketing Board was established in 1933. See Wilmot, "From 'Public Service' to Artificial Insemination," 411–413; Woods, "The Farm as a Clinic."

78. See "Dairy Cattle Exhibition in Emek Izrael," *Cowboy Pages* 1 (1958): 1.

79. "The Cow That Won Two Rewards—Zkufa," *Davar*, December 24, 1937, 3; "8937k of Milk—The Annual Yield of One Cow," *Davar*, November 16, 1937, 6.

80. Levi, *The History of Dairy Bovine in Israel*, 7–8.

81. Israel Palvitch, interview by Nir Mann, Kibbut Ma'abarot, 1996.

82. According to his colleagues, Smaragd was inspired by Jewish tradition, but he disliked both scientists and religious people: "His language was rich and pictorial, and he used expressions from Yiddish and the Bible, of which he was very knowledgeable. Even though he despised religious people . . . and called them pigs." Palvitch, interview by Nir Mann. Moreover, "he was ambivalent towards researchers and men of science. He rejected things he didn't like, and his main criteria was practicality." Professor Ra'anan Volcani, interview by Nir Mann, Rehovot, 1996.

83. Palvitch, interview by Nir Mann. Articles regarding dairy cattle and milk production were frequently published in *Hasade*, a journal established in 1920 by the Hapoel Hatsair workers' movement. The Cattle Breeders Association established its independent magazine, *Meshek Habakar Vehahalav*, in 1952. Dori, *News from the Past, Part 2*, 38.

84. Levi, *The History of Dairy Bovine in Israel*, 7–8.

85. I make use of the herdbook of Kibbutz Kfar-Giladi in order to reconstruct the productive life of the cow Stavit. Novick, "All about Stavit."

86. Quoted in "Will Palestine Become a Center for Cow Breeding?" *Haqiqat al-Amr*, July 12, 1939, 2.

87. "Cows from Palestine for Improving the Cattle Breeds in Iraq and Ceylon," *Haq-iqat al-Amr*, December 16, 1941, 3.

88. I am tempted to interpret the fear of the degeneration of crossbreeds using Ann Stoler's analysis of intimate relationships created between the colonizer and colonized, and the emergence of anxiety and ultimately theories of race when mixing became too prevalent. Ann Laura Stoler, *Carnal Knowledge and Imperial Power: Race and the Intimate in Colonial Rule* (Berkeley: University of California Press, 2010), 62–98.

89. Pinthus, *History of Agricultural Research in Israel*, 333.

90. Dori, *News from the Past, Part 3*, 7, 63–64, 66.

91. In contrast to the European dairy industry, the Israeli one specializes in soft, skim dairy products, with an average low percentage of fat.

92. "Sterility of cows is increasing in settlement. The condition causes heavy economic losses." Quoted in Dr. S. Freund and Dr. A. Rosen, *The Sterility of Cows* (Tel Aviv: Agricultural Experimental Station, Zionist Organization, 1925).

93. Quoted in Levi, *The History of Dairy Bovine in Israel*, 129–130.

94. The committee for the Product of the Land was established in 1934 and revolved around convincing the Jewish population in Palestine to purchase local food. Ayala Plezental, among other scholars, contends that the Product of the Land was not merely about purchasing local products but rather products that were part of the Jewish (versus Arab or German) economy. Plezental, "Milky Way," 134.

95. A promotional poster about the distribution of Hebrew milk in Tel Aviv in the mid-1930s declared, "To the Hebrew Public in Tel-Aviv. With the increase of Hebrew milk in our land . . . the Milkmen Association fulfilled his decision as to distribute MILK FROM THE HEBREW FARMS. . . . [W]e ordered all our members to supply our customers with HEBREW MILK ONLY." Lavon Institute for Labor Research Archives, 252 IV, 12.

96. W. Eilenbogen, "Infertility in Cattle: Reasons and Ways of Treating," *Hasade* 13 (1933): 12.

97. See, for example, S. J. Gilbert, "An Unusual Strain of Brucella Causing Abortion of Cattle in Palestine," *Journal of Comparative Pathology and Therapeutics* 43 (1930): 118–124. Brucellosis severely affected dairy herds in Britain as well and was discussed widely during the period. See Woods, "The Farm as a Clinic."

98. "Our Cowshed in 1944, Kibbutz Mizra'," Kibbutz Mizra' Archives, agricultural sectors, cowshed, 1938–1948.

99. Mordechai Gutman, "How Have We Coped with Infertility in Our Herd?," *Hasade* 20, no.9 (1940): 45. For an analysis of the work done at the experimental station at Kibbutz Gvat, see Hadas Yaron, *Zionist Arabesques: Modern Landscapes, Non-Modern Texts* (Boston: Academic Studies Press, 2010).

100. Saul, "Artificial Insemination," report on the state of the cowshed at Kibbutz Degania-Aleph, 1938, DCBA, 10, 1, 3.

101. During this period, the USSR was the main center for innovation and experimentation with artificial insemination in livestock. For the story of Dr. Henry P. Fox, who arrived in 1934 to Palestine from the USSR with knowledge about artificial insemination practices, see Dori, *News from the Past, Part 3*, 49–51. Interestingly, historian Jenny Leigh Smith argues that in the Soviet Union in 1930 (and as opposed to the United States), artificial insemination along with other "advances in the nascent science of genetics" did not exist and "proved difficult to introduce." The wide circulation in both the German- and English-speaking worlds of Russian Elias Iwanoff's early twentieth-century publications, which deal with the results of his experimentation with artificial insemination in livestock, seems to undermine this assertion. Jenny Leigh Smith, *Works in Progress: Plans and Realities on Soviet Farms, 1930–1963* (New Haven, CT: Yale University Press, 2014), 33; Christina Benninghaus, "Great Expectations: German Debates about Artificial Insemination in Humans around 1912," *Studies in History and Philosophy of Biological and Biomedical Sciences* 38 (2007): 376.

102. Levi, *The History of Dairy Bovine in Israel*, 109.

103. T. Bell, "Preliminary Work on the Artificial Insemination of Live Stock in Palestine," *Empire Journal of Experimental Agriculture* 6 (1938): 188.

104. Dori, *News from the Past, Part 1*, 61–62.

105. "Artificial Insemination—Training Course for Veterinary Doctors," DCBA, 10/2, 1, 2; "Artificial Insemination," lecture to the Jewish veterinary doctors' meeting, DCBA, 10, 2, 1, 2; picture of first class of cattle inseminators, DCBA, Image Collection.

106. Efraim Smaragd to the Soviet Institute for Artificial Impregnation, 1938, DCBA, 10, 1, 2.

107. Dori, *News from the Past, Part 1*, 69.

108. Ra'anan Volcani, "The Beginnings of Artificial Insemination in the South," Cattle Breeders Association Archives, 10, 1, 23–4. In *A Pigeon and a Boy*, Israeli novelist Mei'r Shalev tells a story of a pigeon that carried human sperm during the 1948 war. The sperm, which was sent by a soldier at war, reached his lover after the soldier died in battle. Meir Shalev, *A Pigeon and a Boy: A Novel* (Tel Aviv: Am Oved, 2006).

109. "Sperm of Bulls Arrived from the United-States by Plane," *Al-Hamishmar*, February 10, 1947, 4; report by Ra'anan Volcani to Menahem Tzentner, February 10, 1947, in Dori, *News from the Past, Part 1*, 67. Three years later, Smaragd brought thousands of cows from North America to the State of Israel, establishing the US Holstein cow as the Israeli breed.

110. Ami Neria, *Veterinary Medicine in the Land of Israel: 50 Years of Veterinarian Medicine, 1917–1967* (Tel Aviv: Re'emim Press, 2001), 60.

111. Ruth Bondi, "Reception—For Animal Only," *Davar*, February 3, 1956, 20.

112. "Our Granaries Have Filled with Grain" was written in Hebrew by Pinhas El'ad (Lender) and the music composed by David Zehavi in 1932–1933. I thank Shira Shmuely for reminding me of this song.

113. Deuteronomy 28:1, 28:8. For a look at the relation between god and the product of the land in the Jewish tradition, see Mary Douglas, *Purity and Danger: An Analysis of Concepts of Pollution and Taboo* (New York: Frederich A. Praeger Publishers, 1966), 50.

114. Quoted in Neria, *Veterinary Medicine in the Land of Israel*, 74–75.

115. Zalman Greenberg and Anton Alexander, "Israel Jacob Kligler: The Story of 'A Little Big Man,'" *Korot—The Israeli Journal of the History of Medicine and Science* 21 (2012): 193, 202.

116. Neria, *Veterinary Medicine in the Land of Israel*, 196. According to his colleague Ra'anan Volcani (interview by Nir Mann), Smaragd distrusted scientists, except for "Prof. Saul Adler from the Hebrew University [who] won his full recognition."

117. I heard this argument during a meeting with Arieh Shadar, archivist for the Israeli Breeders Association, who was a milking instructor and inseminator for over fifty years (see figure 4.5), but found little evidence for European-trained human doctors who became veterinarian doctors in Palestine. Two such figures, though, were Arieh Biham (1877–1941) and Zerakh Gilmovsky (1893–1979). See Dori, *News from the Past, Part 2*, 16, 30. A third example was Felix Gad Sumlan (1907–1986) (see chapters 3 and 5), who as a pharmacologist and hormone researcher, gained both DVM and MD degrees prior to immigrating to Palestine in 1933.

118. Such is the case of Bader, who studied medicine in Europe but never graduated. In Palestine/Israel, she worked to simultaneously nurse sick cows in the "sick cowshed" and treat members of the kibbutz in the kibbutz clinic. Berlovitch, *Wounded Bird*. Miriam Baratz was another farmer who took care of her babies and cows concurrently. Smadar Sinai, *Miriam Baratz—Portrait of a Pioneer* (Ramat-Ef'al: Yad Tebenkin, 2002).

119. Neria, *Veterinary Medicine in the Land of Israel*, 221–222.

120. Quoted in Neria, *Veterinary Medicine in the Land of Israel*, 222.

121. Cows were not usually considered Jewish but instead Hebrew. Nevertheless, "Jewish" and "Hebrew" are often used interchangeably in the context of Palestine/Israel. For the sake of relating to the historiography of Zionism and the New Jew, I propose using the New Jewess for both female cows and women.

122. The permeability of reproductive knowledge between humans and animals in Palestine in the 1940s is early in comparison to what has been described in the literature. Adele C. Clarke, for example, writes that "the boundaries between agriculture, medicine and biology were not only porous, but began disappearing for

the reproductive sciences in the early 1970s." Adele C. Clarke, "Reflections on the Reproductive Sciences in Agriculture in the UK and US, ca. 1900–2000+," *Studies in History and Philosophy of Biological and Biomedical Sciences* 38 (2007): 328.

123. Dori, *News from the Past, Part 2*, 66.

124. "Efraim Smaragd," biographical notes, estate of Efraim Smaragd, 2. The success of the new Hebrew cow was so impressive that it finally convinced agriculturalists in the Netherlands, the model country for Jewish dairy cattle management in Palestine, to import Holstein cows to the Netherlands as well. See Theunissen, "Breeding for Nobility or for Production?"

125. Cited in Neria, *Veterinary Medicine in the Land of Israel*, 239.

CHAPTER 5

1. Robert E. Kohler, "Lab History: Reflections," *Isis* 99 (2008), 761–768; Bruno Latour, "Give Me a Laboratory and I Will Raise the World," in *The Science Studies Reader*, ed. Mario Biagioli (New York: Routledge, 1999), 267.

2. For a study of the production of sex hormones at the Organon pharmaceutical company in the Netherlands, see Nelly Oudshoorn, *Beyond the Natural Body: An Archaeology of Sex Hormones* (New York: Routledge, 1994). Sociologist Adele Clarke famously analyzed the rise of the reproductive sciences in the US context. Adele Clarke, *Disciplining Reproduction: Modernity, American Life and the Problem of Sex* (Berkeley: University of California Press, 1998). Naomi Pfeffer wrote about the management of infertility in the United Kingdom and Carl Gemzell's work with human pituitary glands in Sweden. Naomi Pfeffer, *The Stork and the Syringe: A Political History of Reproductive Medicine* (Cambridge, UK: Polity Press, 1993); Naomi Pfeffer, "Pioneers of Infertility Treatment," in *Women and Modern Medicine*, ed. Lawrence Conrad and Ann Hardy (Amsterdam: Editions Rodolpi B. V., 2001). For an exploration of the work of Adolf Butenandt and the German Schering AG pharmaceutical company, see Jean-Paul Gaudillière, "Better Prepared than Synthesized: Adolf Butenandt, Schering AG, and the Transformation of Sex Steroids into Drugs (1930–1946)," *Studies in History and Philosophy of Biological and Biomedical Sciences* 36, no. 4 (2005): 612–644. For a look at scientist Axel Westman and the production of the drug Gonadex by the Swedish pharmaceutical company Leo, see Christer Nordlund, *Hormones of Life: Endocrinology, the Pharmaceutical Industry, and the Dream of a Remedy for Sterility, 1930–1970* (Sagamore Beach, MA: Watson Publishing International, 2011). For an examination of the pharmaceutical production of the pill, see Lara Marks, *Sexual Chemistry: A History of the Contraceptive Pill* (New Haven, CT: Yale University Press, 2001); Gabriela Soto Laveaga, *Jungle Laboratories: Mexican Peasants, National Projects, and the Making of the Pill* (Durham, NC: Duke University Press, 2009).

3. In *Disciplining Reproduction*, Clarke paid attention to the agricultural motivations behind the development of the reproductive sciences. In addition, Clarke

participated in a workshop titled Between the Farm and the Clinic: Agriculture and Reproductive Technology in the Twentieth Century. Held at Cambridge University in 2005, the workshop brought together eminent scholars of reproductive technologies. A collection of essays published through the conference in *Studies in History and Philosophy of Biological and Biomedical Sciences* in 2007 dealt with the emerging reproductive knowledge of animals and humans. See, for example, Sara Wilmot, "From 'Public Service' to Artificial Insemination: Animal Breeding Science and Reproductive Research in Early Twentieth-Century Britain," *Studies in History and Philosophy of Biological and Biomedical Sciences* 38 (2007): 411–441; Abigail Woods, "The Farm as a Clinic: Veterinary Expertise and the Transformation of Dairy Farming, 1930–1950," *Studies in History and Philosophy of Biological and Biomedical Sciences* 38 (2007): 462–487.

4. Soto Laveaga's *Jungle Laboratories*, which revolves around the collection of the Mexican wild yam for the production of the pill, is different in this way, as it pays attention to the socioeconomic context in which the collection campaign happened; it does not discuss users, however. Nordlund's *Hormones of Life* also mentions (but does not analyze) the relation between the production of infertility drugs and the Swedish demography politics of the mid-twentieth century.

5. Dr. Yaski, Hadassah Medical Organization, to the Hebrew University, June 25, 1934, Hebrew University Archives [hereafter HUA], Bernhard Zondek Files, 1; headquarters, Hadassah Medical Organization, to J. L. Magnes, "Reminder (Secret)," December 1, 1934, HUA, Bernhard Zondek Files, 1.

6. Zondek was invited to Sweden and worked at the biochemical laboratory at Stockholm University for about a year between his expulsion from Berlin in 1933 and move to Jerusalem in 1934. Grant applications at the Rockefeller Foundation's archives indicate that he continued his sex hormone research there.

7. Professor Yehuda Leib Magnes, Hebrew University of Jerusalem, to Professor Bernhard Zondek, December 4, 1934, National Library of Israel Archives [hereafter NLIA], NLIA/Hermann Zondek Archive, Bernhard Zondek Files, Arc 4°1674, 150, 1. This document also exists among Zondek's files at the HUA.

8. Zondek identified two types of gonadotropins produced in the pituitary gland, and named them Prolan A and Prolan B. These terms were adopted by the global scientific community and used for about two decades, but later replaced by follicle-stimulating hormone and luteinizing hormone, respectively. For a discussion of the contemporary nomenclature, see Bernhard Zondek and Felix Sulman, *The Antigonadotropic Factor, with Consideration of the Antihormone Problem* (Baltimore: Williams and Wilkins Company, 1942), 3.

9. Adele Clarke, "Research Materials and Reproductive Science in the United States, 1910–1940," in *Physiology in the American Context 1850–1940*, ed. Gerald L. Geison (New York: Springer, 1987), 331; Marius Tausk, *Organon: The Story of an Unusual Pharmaceutical Enterprise* (Oss: Akzo Pharma bv, 1984), 24–26.

10. Bernhard Zondek, "Studies of the Mechanism of the Female Genital System," *Fertility and Sterility* 10, no. 1 (1959): 7. "Tropic," for controlling or influencing, and hence "gonadotropic" hormones are those secreted in the pituitary, and influence the hormonal activity in the gonads, the reproductive organs.

11. Zondek, "Studies of the Mechanism of the Female Genital System," 7. This description of "motor" and "conductor of symphony" was then repeated by many of Zondek's colleagues. See, for example, Tausk, *Organon*, 30.

12. Bernhard Zondek, "Mass Excretion of Oestrogenic Hormone in the Urine of Stallions," *Nature* (February 10, 1934): 209–210; Bernhard Zondek, "Oestrogenic Hormone in the Urine of the Stallion," *Nature* (March 31, 1934): 494; Tausk, *Organon*, 26. Biochemist Marius Tausk (*Organon*, 26) of the pharmaceutical company Organon notes that "this perplexing discovery [of 'female' sex hormones in stallions] . . . was of physiological (theoretical) interest and of no use for the manufacture." Oudshoorn (*Beyond the Natural Body*, 25–26) analyzes how the rise of endocrinology, and Zondek's discoveries in particular, diminished the believed dichotomy between "female" and "male" hormones, and hence the perceived dichotomy of female and male bodies.

13. The first two international conferences for the standardization of hormones were held in 1932 and 1935 in London. See Oudshoorn, *Beyond the Natural Body*, 46; Nordlund, *Hormones of Life*, 35–36. Zondek was invited by the League of Nations to participate in the standardization of sex hormones, and complained to the Hebrew University that he did not have enough funds for the travel. Bernhard Zondek to the management of the Hebrew University, January 13, 1938, HUA, Bernhard Zondek Files, 1.

14. In the original version of the test, the mouse was not merely a "guinea pig" but instead was used as disposable laboratory equipment and sacrificed in the process. The rabbit and frog tests (or more accurately, "toad test") were actually an improvement on the original A-Z test, using different animals that enabled easier detection of pregnancy, and in the case of the toad, without sacrificing the animal in the process. On the scale of the use of the A-Z test, see Jesse Olszynko-Gryn, "The Demand for Pregnancy Testing: The Aschheim-Zondek Reaction, Diagnostic Versatility, and Laboratory Services in 1930s Britain," *Studies in History and Philosophy of Biological and Biomedical Sciences* 47 (2014): 238. See also Jesse Olszynko-Gryn, *A Woman's Right to Know: Pregnancy Testing in Twentieth-Century Britain* (Cambridge, MA: MIT Press, 2023).

15. Dr. Bernard J. Brent, Montclair, New Jersey, United States, to Professor Bernhard Zondek, Jerusalem, Israel, July 12, 1951, NLIA, Arc4°1674, 157. When sending the letter, Brent worked at the Roche-Organon pharmaceutical laboratories in New Jersey. Zondek was an adviser in the mid-1920s for Organon in the Netherlands, where he was able to promote urine as the ideal source for sex hormones. See Tausk, *Organon*, 23–27. The letter from Brent together with numerous other comments in

Zondek's publications and letters indicate that he continued to be in touch with Organon throughout the years.

16. There is a long tradition of using urine as a truth-searching resource. A main example is that of uroscopy, which was the most important diagnostic method in Europe in the High Middle Ages, but began declining in the early sixteen century. See Michael Stolberg, "The Decline of Uroscopy in Early Modern Learned Medicine, 1500–1650," *Early Science and Medicine* 12 (2007): 313–336. Urine was also a basic material for alchemist investigations, ultimately leading to the discovery of phosphate by Hennig Brand of Hamburg in the 1660s.

17. Clarke, *Disciplining Reproduction*, 291; Pfeffer, "Pioneers of Infertility Treatment," 253; Nordlund, *Hormones of Life*, 40.

18. Clarke, "Research Materials and Reproductive Science," 323–341; Oudshoorn, *Beyond the Natural Body*, 48.

19. After his expulsion, Zondek asked his community to support him. His archival material is full of recommendation letters written by a variety of colleagues from around the world emphasizing the significance of the scientific work he accomplished in Germany. In an interview in 1966, Zondek mentioned that his British colleagues wrote a collective letter of support, which was published in the *Times* in 1933. Michael Finkelstein, "Professor Bernhard Zondek—An Interview," *Journal of Reproduction and Fertility* 12 (1966): 11.

20. Dr. Yaski, Hadassah Medical Organization, to the Hebrew University, June 25, 1934, HUA, Bernhard Zondek Files, 1; Dr. Yaski to Dr. Nathan Ratnoff, 1934, HUA, Bernhard Zondek Files, 1; Hebrew University of Jerusalem, Obituary—Bernhard Zondek, "For Immediate Release," HUA, Bernhard Zondek Files, 2; Professor David Amiran, "Prof. Bernhard Zondek—Obituary," HUA, Bernhard Zondek Files, 3.

21. Marcia Gitlin, "A Famous Scientist Comes Home," news from the Canadian Friends of the Hebrew University, September 3, 1951, 2, NLIA/Hermann Zondek Archive, Bernhard Zondek Files, Arc 4°1674, 188.

22. Daniel P. O'Brien, Rockefeller Foundation, New York, to Bernhard Zondek, Switzerland, October 15, 1934, HUA, Bernhard Zondek Files, 1; Daniel P. O'Brien to J. L. Magnes, January 28, 1935, HUA, Bernhard Zondek Files, 1. For a detailed analysis of the financial support supplied by the Rockefeller Foundation through the American National Research Council Committee for Research in Problems of Sex, see Clarke, *Disciplining Reproduction*, 90–120.

23. Nordlund, *Hormones of Life*, 38. For numerous reference letters sent in late May 1947 by leading fertility experts around the world in support of Zondek, probably to support his nomination for the Nobel Prize, see, for example, file no. AK, Central Zionist Archives [hereafter CZA], 576, 1. Among those letters were George W. Corner, director of the American National Research Council's Committee for the Research in Problems of Sex, and US biochemist Edward Adelbert Doisy, codeveloper of the Allen-Doisy test (1923), the precursor to the A-Z test.

24. For correspondence regarding Sulman's employment, see Professor Bernhard Zondek to Dr. Ben-David, Secretariat, Hebrew University of Jerusalem, April 25, 1937, HUA, Felix Sulman Files, until 1953, 1; Dr. Felix Sulman to the management of the Hebrew University of Jerusalem, July 10, 1938, HUA, Felix Sulman Files, until 1953, 1. On the approval of the budget for his employment, see "Minutes of Meeting between Dr. Senator, Prof. Zondek, Mr. Schneorson, and Dr. Ben-David," August 12, 1940, HUA, Felix Sulman Files, until 1953, 1. In reality, Sulman officially became a senior assistant only in late 1944. See M. Ben-David to Dr. F. Sulman, January 4, 1945, HUA, Felix Sulman Files, until 1953, 1. The work done at the lab was continuously scrutinized over the years, as indicated by a letter Zondek wrote to neurologist Israel S. Wechsler in 1951. In this letter, Zondek notes that he read a report composed by a group of experts, including Wechsler, recommending that the lab should be closed and remarking that under the contemporary circumstances, it considered hormone research "to be something of luxury." Zondek replied to those allegations by arguing that "modern Gynecology and Obstetrics in my opinion [are] not workable without such laboratory." Bernhard Zondek to Dr. Wechsler, August 17, 1951, NLIA, Arc4°1674, 148.

25. Sulman, born in Berlin, gained both his DVM (1930) and MD (1933) degrees from Berlin University (now Humboldt University of Berlin), and worked as an assistant in the Prussian Institute for Hygiene and Immunology in Dahlem, Berlin, during 1930–1932. He immigrated to Palestine in 1933 and started working in Zondek's lab in 1934. See Felix Sulman to the management of the Bialik Institute, December 5, 1948, CZA, S83, 277, 2.

26. Clarke, "Research Materials and Reproductive Science," 335. The "small-to-big" equation is, of course, a simplification. In the field of sex hormones, for example, primates were studied as well and whales were chosen to be suppliers of "male sex hormones" due to the size of their testes. Oudshoorn, *Beyond the Natural Body*, 67. Primates were especially important as research subjects as part of the attempts to understand menstruation, as they are the only nonhuman menstruating species.

27. See, for example, Felix Gad Sulman, *Hormones: Textbook for Hormone Theory and Use* (Tel Aviv: Hasade Library, 1962). This type of translation is particularly note-worthy due to the fact the Zondek stopped publishing in German in the mid-1930s, as did other European scientists with the rise of the Nazi rule.

28. *Endocrinology, Clinical Endocrinology,* and *Acta Endocrinologica* are some examples.

29. On pregnancy tests, see, for example, Felix Gad Sulman and Edith Sulman, "Pregnancy Test with the Male Toad," *An International Journal of Obstetrics and Gyne-cology* 56, no. 6 (1949): 1014–1017; Professor Zondek and Dr. Sulman, "Preparation of Oestrogenic Hormone from the Urine of Pregnant Women and Equines," research proposal submitted to the Board for Scientific and Industrial Research, government of Palestine, Israel State Archives [hereafter ISA], M-52, 1881. In late 1934, for the purpose of applying for the Rockefeller grant, Zondek explicitly stated what he was

planning to research in the lab in Jerusalem: "1. The relationship of cancer and hormones; 2, Hormones and their action upon the maturity of the egg, 3. The relationship of the Manura Gland and Sexual Function." Bernhard Zondek to Dr. O'Brien, October 31, 1934, HUA, Bernhard Zondek Files, 1. On the purification of hormones, see Bernhard Zondek, "On the Mechanism of the Action of Gonadotropin from Pregnancy Urine," *Journal of Endocrinology* 2 (1940): 12–20; Bernhard Zondek and Rivka Black, "Is Normal Human Urine Toxic?," *Proceedings of the Society for Experimental Biology and Medicine* 61 (1946): 140–142.

30. Zondek and Sulman, *The Antigonadotropic Factor*; Bernhard Zondek and Y. M. Bromberg, "Testing and Treatment of Sensitivity to Endogenous Hormones," research proposal submitted to the Board for Scientific and Industrial Research, government of Palestine, ISA, M-41, 1881; Bernhard Zondek and Y. M. Bromberg, "Endocrine Allergy: Clinical Reaction of Allergy to Endogenous Hormones and Their Treatment," *Journal of Obstetrics and Gynecology of the British Empire* 54, no. 1 (1947): 1–19.

31. Nordlund, *Hormones of Life*, 171.

32. Zondek and Sulman, *The Antigonadotropic Factor*, 147. See also Bernhard Zondek and Felix Sulman, "The Antigonadotropic Factor: Species Specificity and Organ Specificity," *Proceedings of the Society for Experimental Biology and Medicine* 36 (1937): 712–717. Clarke (*Disciplining Reproduction*, 192, 224) argues that in the 1930s, the study of antihormone was considered by some to hold a promise for the development of contraceptives. Zondek and Sulman indeed commented in 1940 that antihormones might be used to stop pregnancies, but that such treatment was yet to prove itself. See Bernhard Zondek and Felix Sulman, "The Role of Sex Hormones in the Sex Mechanism of the Animals (Lecture Given at the Jewish Organization of Veterinary Services Annual Conference)," *Alon Histadrut Harofim Haveterinarim Beeretz Israel* 5 (1940): 8.

33. Zondek and Black, "Is Normal Human Urine Toxic?"

34. See, for example, Dr. S. Freund and Dr. A. Rosen, *The Sterility of Cows* (Tel Aviv: Agricultural Experimental Station, Zionist Organization, 1925); W. Eilenbogen, "Infertility in Cattle: Reasons and Ways of Treating," *Hasade* 13 (1933): 12–15.

35. Ami Neria, *Veterinary Medicine in the Land of Israel: 50 Years of Veterinarian Medicine, 1917–1967* (Tel Aviv: Re'emim Press, 2001), 74–75.

36. Zondek and Sulman, "The Role of Sex Hormones in the Sex Mechanism of the Animals," 9. A copy of this lecture is also found in the National Library, Zondek's collection at the Zionist Archives, and the Breeders Association Archives.

37. For an acknowledgment of sending books and articles, see Bernhard Zondek in Jerusalem to Aharon Harari on Kibbutz Merhavia, May 3, 1941, Kibbutz Merhavia Archives [hereafter KMA], Aharon Harari Archive, 2.8.

38. This indeed happened less than twenty years later. Most of the herds of sheep, which existed on the great majority of the kibbutzim from the 1910s to 1920s, were gradually terminated in the late 1950s and early 1960s. See chapter 3.

39. Sulman and Harari developed personal ties over the years, and used to visit each other with their families during holidays. Sulman and his wife, Edith, would offer their Jerusalem house during hot summer days, and Harari would invite Sulman's family to relax in the countryside. See correspondence between Sulman and Harari in Harari's files in the Kibbutz Merhavia Archives.

40. In their correspondence, Harari and Sulman often mentioned the use of buses to transport serum, urine, or plant samples as well as complain about the limitations of Egged, the Jewish transportation company. To read about other interesting means of transporting urine to labs for examination, see the case of the 1930s' British mailing systems, which transported many thousands of packages containing urine of women to the Edinburgh laboratory for detection of pregnancy using the A-Z test. Olszynko-Gryn, "The Demand for Pregnancy Testing," 236.

41. Zondek and Sulman apparently managed to speed up the results of the pregnancy test for mares by using a method that entailed the sacrifice of four rats. Bernhard Zondek and Felix Sulman, "A Twenty-Four Hour Pregnancy Test for Equines," *Nature* 455 (March 10, 1945): 320.

42. Prolans are the names that Zondek gave to gonadotropins. See note 8.

43. Dr. Felix Sulman to Herr Harari, May 3, 1946, KMA, Aharon Harari Archive, Sheep: Correspondence with Academia and the Guidance Department at Rehovot.

44. Felix Sulman, Hebrew University of Jerusalem, to Aharon Harari, Kibbutz Merhavia, May 29, 1944, KMA, Aharon Harari Archive, Sheep: Correspondence with Academia and the Guidance Department at Rehovot.

45. "I.U." refers to the standardized hormone unit, or "international unit," the result of the international hormone conferences that took place in the 1930s.

46. Dr. Felix Sulman to Herr Harari, April 18, 1945, KMA, Aharon Harari Archive, Sheep: Correspondence with Academia and the Guidance Department at Rehovot.

47. Felix Sulman to B. Gila'di, Organization of Labor, Carrier, and Riding Animals, Nahalal, May 29, 1944, KMA, Aharon Harari Archive, Sheep: Correspondence with Academia and the Guidance Department at Rehovot. The names of the mares mentioned here are interesting, and are probably Sulman's symbolic choice rather than a reference to real horses. The first—Fatmeh—being a female Arab name, and the second—Sonia—a non-Hebrew name, somewhat common among eastern European Jewish female settlers, seem far from random. It appears that Sulman chose these names to imply that there were problems with the cooperation of both Jewish and Palestinian farmers. If this is indeed the case, this is the only existing reference indicating that Palestinian people were incorporated into the efforts to deal with infertility in that period.

48. Harari corresponded with thirty-eight different agricultural settlements about sheep management during the years 1939–1962, with many on the issue of hormone therapy. See KMA, Aharon Harari Archive, Sheep: Correspondence with different settlements.

49. Shepherds of Kibbutz Kfar-Sald to Avraham [by mistake] Harari, Rehovot, August 8, 1945, KMA, Aharon Harari Archive, Sheep: Correspondence with different settlements.

50. Guidance Department, Agriculture Experimental Station, Rehovot, to Aharon Harari, re: report on your experiments with injecting hormones to sheep, April 7, 1946, and August 6, 1946, KMA, Aharon Harari Archive, Sheep: Correspondence with Academia and the Guidance Department at Rehovot.

51. Aharon Harari, "Report on the Actions Taken to Increase Fertility among the Sheep in the 1945 Season," KMA, Aharon Harari Archive, Sheep: Articles and Reports, 4–5.

52. Nordlund, *Hormones of Life*, 70.

53. Quoted in Nordlund, *Hormones of Life*, 70–71.

54. Zondek and Sulman, "A Twenty-Four Hour Pregnancy Test for Equines."

55. See, for example, Bernhard Zondek, "Impairment of Anterior Pituitary Functions by Follicular Hormone," *Lancet* (October 10, 1936): 842–847. On page 842, he noted, "My starting-point [for the study] was a clinical observation."

56. Oudshoorn, *Beyond the Natural Body*, 117, 135; Clarke, *Disciplining Reproduction*, 145.

57. Zondek, "Impairment of Anterior Pituitary Functions by Follicular Hormone," 842.

58. "Sterility in Women Becoming a Serious Problem: Professor Zondek," *Winnipeg Tribune*, September 3, 1952.

59. Finkelstein, "Professor Bernhard Zondek," 3–19.

60. Bernhard Zondek, *Estrous Hormone and a Colored Matter Resembling Vitamin B2 in the Dead Sea* (Jerusalem: Magnes Books, 1938), 1–5.

61. Bernhard Zondek, "Oestrogenic Substances in the Dead Sea," *Nature* 140, no. 3536 (1937): 240.

62. B. Zondek, "Study of the Occurrence of Sex Hormones in Palestinian Plants," research proposal submitted to the Board for Scientific and Industrial Research, government of Palestine, ISA, M-46, 1881.

63. B. Zondek, "Study of Hyalonuridase in the Sperm of Animals and Man and Its Bearing to Seasonal Sterility," research proposal submitted to the Board for Scientific and Industrial Research, government of Palestine, August 3, 1947, ISA, M-46, 1881.

64. Bernhard Zondek, "Studies of the Mechanism of the Female Genital System," *Fertility and Sterility* 10, no. 1 (1959): 12.

65. Grace M. Crowfoot and Louise Baldensperger, *From Cedar to Hyssop: A Study in the Folklore of Plants in Palestine* (London: Sheldon Press, 1932), 106–126. Crowfoot and Baldensperger explain the long tradition of using the mandrake as a love charm, among both Jewish and Arab communities in Palestine and Syria. On pages 114–115, for example, they tell of a Jewish childless man in Chicago who in 1891 received a mandrake from Jerusalem as a talisman for fertility.

66. Felix Sulman to Aharon Harari, March 16, 1952, KMA, Aharon Harari Archive, Sheep. As part of their correspondence regarding the mandrake samples, Sulman writes to Harari, "Many thanks for the juices [of the Mandrake fruit] that arrived in the meantime. I began the examination. The endocrinology test came as a god of heaven to our wrecked house, and I thank you from the bottom of my heart."

67. Felix Sulman and Gavriella Tietz, "The Sex-Hormones of the Mandrake," *Harefua—Journal of the Palestine Jewish Medical Association* 23, no. 7 (1942): 1–4. See also Zondek's review of Sulman's findings in Zondek, "Study of the Occurrence of Sex Hormones in Palestinian Plants."

68. Felix Sulman to Aharon Harari, October 9, 1945, date unknown, KMA, Aharon Harari Archive, Correspondence with Academia and the Guidance Department at Rehovot; Felix Sulman to Aharon Harari, November 2, 1958, date unknown, KMA, Aharon Harari Archive, Correspondence with Academia and the Guidance Department at Rehovot; Aharon Harari and Felix Gad Sulman, "Alignment of the Estrous Cycle in Sheep by Light," date unknown, KMA, Aharon Harari Archive, Correspondence with Academia and the Guidance Department at Rehovot.

69. Felix Gad Sulman. *Hypothalamic Control of Lactation: Monographs on Endocrinology* (London: William Heinemann Medical Books, 1970), 2, 195–197.

70. Call for participants in a committee on climatic effects on domestic animals, Meteorological Research Council, by Felix Sulman, September 27, 1945, HUA, Felix Sulman, until 1953, 1.

71. Archival materials show that Sulman received a $30,000 grant from the US Office of Aerospace Research to investigate the effect and prevention of climatic heat stress. See HUA, Felix Sulman Files, 1954–1968, 2.

72. Felix Gad Sulman, *Health, Weather and Climate* (Basel: S. Krager, 1976), 5.

73. With Zondek's approval, Sulman requested to work as an instructor in the Department of Applied Pharmacology already in 1950. By 1955, he was appointed a lecturer in that department. See Felix Sulman to the management at the Hebrew University, May 18, 1950, HUA, Felix Sulman Files, I; "Excerpt of the Advisory Board Meeting," January 5, 1955, HUA, Felix Sulman Files, I.

74. Bernhard Zondek to Professor A. L. Olitzky, dean of the medical school, November 9, 1961, 1–2, HUA, Bernhard Zondek Files, III.

75. Bernhard Zondek, "Studies of the Mechanism of the Female Genital System," *Fertility and Sterility* 10, no. 1 (1959): 12.

76. Bernhard Zondek to the management of the Hebrew University, June 10, 1965, HUA, Hormones and Gynecology, 1965; "Prof. Bernhard Zondek Died Two Days Ago in New-York," *Ma'ariv*, November 10, 1966, 2.

77. Zondek, "Studies of the Mechanism of the Female Genital System," 12–13.

78. Marcia Gitlin, "Bernhard Zondek—A Great Doctor in Israel," *Zionist Record*, September 21, 1951; university reporter, "An Award for a Praised Researcher for His Scientific Achievements," *Davar*, April 24, 1957; Bernhard Zondek to the Territories Committee, July 13, 1964, HUA, Hormones and Gynecology, 1964; committee's reply, November 10, 1964, HUA, Hormones and Gynecology, 1964; Michael Finkelstein to A. Shertz, August 2, 1965, HUA, Hormones and Gynecology, 1964. Zondek also received a generous grant from the Population Council of the Rockefeller Institute to conduct this study. See "Bernhard Zondek," excerpt from *News Magazine* 16, October 18, 1960, HUA, Bernhard Zondek Files, III.

79. Bernhard Zondek to Professor Amiran, president, "Urgent," May 6, 1966, HUA, Hormones and Gynecology, 1966.

80. "Prof. Bernhard Zondek Died Two Days Ago in New-York," 2.

81. Oudshoorn, *Beyond the Natural Body*, 88.

82. In 1941, Zondek wrote to the university administration explaining that he was asked to produce sex hormones from the urine of pregnant women hospitalized at the Hadassah Medical Center since there was a shortage of sex hormone supply due to the war. The university refused. Bernhard Zondek to Dr. Senator, November 7, 1941, HUA, Bernhard Zondek Files, I; Senator's reply, November 17, 1941, HUA, Bernhard Zondek Files, I. In 1961, Sulman stated that he worked for all existing pharmaceutical companies in Israel, and listed them all. Felix Sulman to the Employment Committee, Hebrew University of Jerusalem, January 16, 1961, HUA, Felix Sulman Archive. Furthermore, as indicated by many of his publications, Zondek maintained his connections with various pharmaceutical companies around the world, often thanking them for supplying biological materials or conducting experiments on materials he was interested in. In 1938, for example, he thanked Marius Tausk of Organon for helping him to examine the hormonal potential of the Dead Sea. Zondek, *Estrous Hormone and a Colored Matter Resembling Vitamin B2 in the Dead Sea*, 4.

83. Bruno Lunenfeld, "Historical Perspectives in Gonadotrophin Therapy," *Human Reproduction Update* 10, no. 6 (2004): 457.

84. Tikva Weinstock, "Miracle Drug for Sterile Women—In Elderly Homes," *Ma'ariv*, February 1, 1965, 9.

85. Neri Livneh, "My Million Children—An Interview with Bruno Lunenfeld," *Ha'aretz*, May 29, 2002.

CONCLUSION

1. G. S. Emanuel, chief veterinary officer, to deputy director, Soil Conservation Board, March 2, 1941, Israel State Archives [hereafter ISA], Mem-13, 5109.

2. "Production of Honey in Eretz-Israel," report by Israel Robert Blum to the British government in Palestine, September 15, 1936, Central Zionist Archives, S90, 793.

3. Eitan Bar-Yosef, *Villa in the Jungle: Africa in Israeli Culture* (Tel Aviv: Hakibbutz Hameuchad Publishing House, 2013), 32–36.

4. Honaida Ghanim, "What Is the Color of the Arab? A Critical Look at Color Play," in *Racism in Israel*, ed. Yehuda Shenhav and Yossi Yona (Tel Aviv: Hakibbutz Hameuchad Publishing House, 2008), 92. On Zionism and racism against Palestinians, see also Joseph A. Massad, *The Persistence of the Palestinian Question: Essay on Zionism and the Palestinians* (London: Routledge, 2006).

5. William Cronon, *Nature's Metropolis: Chicago and the Great West* (New York: W. W. Norton and Company, 1991), 214–215.

6. *Sallah Shabati*, directed by Ephraim Kishon, produced by Menahem Golan, 1964.

7. Ghassan Kanafani, "The Land of Sad Oranges," in *Men in the Sun and Other Palestinian Stories*, trans. Hilary Kilpatrick (Boulder, CO: Lynne Rienner Publishers, 1999), 75–81. Nasser Abufarha, "Land of Symbols: Cactus, Poppies, Orange and Olive Trees in Palestine," *Identities: Global Studies in Culture and Power* 15 (2008): 343–368.

8. Pierre Bourdieu, *Distinction: A Social Critique of the Judgment of Taste* (Cambridge, MA: Harvard University Press, 1984).

9. Yael Raviv, "Falafel: A National Icon," *Gastronomica: The Journal of Critical Food Studies* 3, no. 3 (2003): 20–25; Dafna Hirsch, "'Hummus Is Best When It Is Fresh and Made by Arabs': The Gourmetization of Hummus in Israel and the Return of the Repressed Arab," *American Ethnologist* 38 (2011): 617–630; Dafna Hirsch and Ofra Tene, "Hummus: The Making of an Israeli Culinary Cult," *Journal of Consumer Culture* 13, no. 1 (2013): 25–45.

10. For an example illustrating the dependence of West Bank Palestinians on the Israeli dairy industry, see *The Wanted 18*, a 2014 stop-motion animation film that re-creates a true story from the first Palestinian Intifada. The film takes place in the town of Beit Sahour and deals with Israeli Defense Forces' pursuit of eighteen cows who were bought from an Israeli kibbutz in order to enable an independent milk supply for the people of the town. The film depicts the efforts to hide the cows from the Israeli authorities, defining the project as "a true story of bovine resistance." *The Wanted 18*, directed by Amer Shomali and Paul Cowan, 2014.

11. Shmuel Stoller, *The Return of the Date Palm to the Land of Israel: Ben-Zion Israeli's Journeys in Iraq, Iran, and Egypt* (Tel Aviv: Hakibbutz Hameuchad Publishing House, 1977).

12. Director of agriculture and forests, government of Palestine, to chief secretary, "Date Cultivation," November 13, 1933, ISA, Mem-3, 4304.

13. Memories of Shmuel Stoller of the Kinneret group, May 27, 1976, Archive of Beit-Gan (Yavne'el), 2, 9, http://www.yavneel.org.il/site/files/file_1326.PDF.

14. Judy Siegel-Itzkovitch, "Medicinal Date Palm from Oldest Known Seed Planted," *Jerusalem Post*, November 25, 2011, http://www.jpost.com/Health/Article.aspx?id =246956. The record was broken in 2012 as a team of Russian scientists managed to generate plants from the fruit of an arctic flower that supposedly died thirty-two thousand years ago. Nicholas Wade, "Dead for 32,000 Years, an Arctic Plant Is Revived," *New York Times*, February 12, 2012.

15. Ofri Ilani, "2,000-Year-Old Date Seed Grows in the Arava," *Ha'aretz*, February 15, 2007, http://www.haaretz.com/print-edition/news/2-000-year-old-date-seed-grows-in -the-arava-1.213054.

16. Sarah Sallon and Elaine Solowey, "Germination, Genetics, and Growth of an Ancient Date Seed," *Science* 320 (2008): 1464. Sallon directs the Louis L. Borek Natural Medicine Research Center at Hadassah Medical Center, a major supporter of the sprouting project (see figure C.3). According to Solowey, Sallon was the one who convinced the Israeli Antiquity Authority to supply Solowey with three date seeds. Interview with Elaine Solowey, Kibbutz Ketura, February 16, 2012. As Sallon noted in various interviews, the belief that ancient dates were used for medicinal purposes intersected with her interest in the medicinal qualities of plants, motivating her to initiate the project.

17. Quoted in Siegel-Itzkovitch, "Medicinal Date Palm from Oldest Known Seed Planted."

18. Quoted in Ilani, "2,000-Year-Old Date Seed Grows in the Arava."

19. Steven Erlanger, "After 2,000 Years, a Seed from Ancient Judea Sprouts," *New York Times*, June 12, 2005.

20. The "Green Line" refers to the armistice lines between the newly established State of Israel and its neighbors in 1949. The term *Hitnahaluyot* ("settlements") commonly refers to Jewish settlements that were established after the 1967 war "outside" these lines and within the Palestinian territories.

21. For data regarding the Jewish growth of palm trees in the West Bank and support of Christian organizations, see Dror Etkes, *Israeli Settler Agriculture as a Means of Land Takeover in the West Bank* (Jerusalem: Kerem Navot, 2013), 79–82. This report is based on an analysis of fifteen years of GIS data collected by the Israeli Civil Aministration, the governing body operating in the West Bank.

22. Interview with Solowey.

23. Nir Hasson, "In an Attempt to Taste the Ancient Fruit, Researchers Germinated Six 2,000-Year-Old Date Seeds," *Ha'aretz*, February 6, 2020, https://www.haaretz .co.il/science/archeology/1.8501814.

24. Sarah Sallon, Emira Cherif, Nathalie Chabrillange, Elaine Solowey, Muriel Gros-Balthazard, Sarah Ivor-Ra, Jean-Frédéric Terral, et al., "Origins and Insights into the Historic Judean Date Palm Based on Genetic Analysis of Germinated Ancient Seeds and Morphometric Studies," *Science Advances* 6, no. 6 (2020): 6.

25. Quoted in Yisha'ayahu Aviam, "The World Congress for Fertility and Sterility Increasing the Prestige of Science in Israel," *Ma'ariv*, May 29, 1968.

26. Ya'acov Atzmon, "With Retirement," *Meshek Habakar Vehahalav* (1977): 3–8; "More than Two Hundred Cows to Iran—Israel Is the Biggest Cattle Supplier to That Country," *Davar*, December 4, 1967, 3.

27. Maxximilk—Much More Milk, http://www.maxximilk.com/index.php?tab=home. Another Israel agritech company is Afimilk, which offers electronic meters, behavior sensors, and other computerized systems to dairy farms around the world. See "Automating Dairy Farms," Afimilk, http://www.afimilk.com.

BIBLIOGRAPHY

"About Cattle Tenders." *Al-Difa'*, March 4, 1945, 3.

Abu El-Haj, Nadia. *Facts on the Ground: Archaeological Practice and Territorial Self-Fashioning in Israeli Society*. Chicago: University of Chicago Press, 2001.

Abu El-Haj, Nadia. *The Genealogical Science: The Search for Jewish Origins and the Politics of Epistemology*. Chicago: University of Chicago Press, 2012.

Abufarha, Nasser. "Land of Symbols: Cactus, Poppies, Orange and Olive Trees in Palestine." *Identities: Global Studies in Culture and Power* 15 (2008): 343–368.

Abu-Lughod, Leila. *Veiled Sentiments: Honor and Poetry in a Bedouin Society*. Berkeley: University of California Press, 1986.

Abu-Rabia, Aref. *The Negev Bedouin and Livestock Rearing: Social, Economic, and Political Aspects*. Oxford: Berg Publishers, 1994.

Adas, Michael. *Machines as the Measure of Man: Science, Technology, and the Ideologies of Western Dominance*. Ithaca, NY: Cornell University Press, 1990.

Advanced Cow Cooling Systems Co. and Hachaklait Veterinary Services Ltd. "Management of Heat Stress to Improve Fertility in Dairy Cows in Israel." *Journal of Reproduction and Development* 56 (2010): 35–51.

Agnon, Shmuel Yosef. *The Tale of the Goat*. Jerusalem: Hagina Publishers, 1925.

Agrawal, Arun. *Environmentality: Technologies of Government and the Making of Subjects*. Durham, NC: Duke University Press, 2005.

Aguilar-Rodríguez, Sandra. "Nutrition and Modernity: Milk Consumption in the 1940s and 1950s Mexico." *Radical History Review* 110 (2011): 36–58.

Alatout, Samer. "Bringing Abundance into Environmental Politics: Constructing a Zionist Network of Water Abundance, Immigration, and Colonization." *Social Studies of Science* 39, no. 1 (2009): 363–394.

Aleichem, Sholem. *The Bewitched Tailor*. Moscow: Foreign Languages Publication House, 1957.

Alexander, Jennifer. "Radically Religious: Ecumenical Roots of the Critique of Technological Society." In *Jacques Ellul and the Technological Society in the 21st Century*, edited by Helena M. Jerónimo, José Garcia, and Carl Mitcham, 191–203. Dordrecht: Springer, 2013.

"Al-Farouk Receives the Son of Al Saud in the City of Suez: A Herd of Goats and Camels for Egypt's Great Visitor." *Filastin*, December 16, 1945, 1.

Algazi, Gadi. "From Gir Forest to Um-Hiran: Comment on Colonial Nature and Its Conservators." *Teoria Vebikoret* 37 (2010): 232–253.

Algazi, Yosef. "Ransoming Captive Camels." *Ha'aretz*, October 14, 1994.

Allen, Henry. "The Holy Land Bees." *American Bee Journal* 20, no. 37 (1884): 586.

Almog, Oz. *The Sabra: The Creation of the New Jew*. Translated by Haim Watzman. Berkeley: University of California Press, 2000.

Al-Nuwayrī, Shihāb al-Dīn. *The Ultimate Ambition in the Arts of Erudition: A Compendium of Knowledge from the Classical Islamic World*. New York: Penguin Books, 2016.

Al-Salim, Farid. *Palestine and the Decline of the Ottoman Empire: Modernization and the Path to Palestinian Statehood*. London: Bloomsbury, 2015.

Alterman, Natan. "The Picture of the Cow 'Stavit.'" *Davar: Hatur Hashvi'i*, August 4, 1950.

Amar, Zohar, and Yaron Serri. "When Did the Water Buffalos Arrive to the Water Landscapes of the Land of Israel?" *Katedra* 117 (2005): 63–70.

Amara, Ahmad, Ismael Abu-Saad, and Oren Yiftachel, eds. *Indigenous (In)Justice: Human Rights Law and Bedouin Arabs in the Naqab/Negev*. Cambridge, MA: Harvard University Press, 2012.

Amiur, Hezi. *Mixed Farm and Smallholding in Zionist Thought*. Jerusalem: Zalman Shazar Center, 2016.

Anderson, Virginia D. *Creatures of Empire: How Domestic Animals Transformed Early America*. Oxford: Oxford University Press, 2004.

Andrews, Thomas G. *Killing for Coal: America's Deadliest Labor War*. Cambridge, MA: Harvard University Press, 2008.

Anker, Peder. *Imperial Ecology: Environmental Order in the British Empire, 1895–1945*. Cambridge, MA: Harvard University Press, 2001.

"Apiary Interests in Palestine." *Geneva Daily Times*, May 5, 1900.

Armbruster, Ludwig. "The Bee in the Orient II: Bible and Bee." *Archiv für Bienenkunde* 7, no. 1 (1932): 1–43.

Arnold, David. *The Tropics and the Traveling Gaze: India, Landscape and Science, 1800–1856*. Seattle: University of Washington Press, 2006.

Arnon-Ohana, Yuval. *Peasants in the Arab Revolt in the Land of Israel, 1936–1939.* Tel Aviv: Shiloah Institute, Tel Aviv University, 1978.

Arraf, Shukri. *The Sources of the Palestinian Economy from the Earliest Periods and until 1948.* Ma'aliya: Dar El Amaq, 1997.

Ashkenazy, Eli. "The Water Buffalos Are Coming Back: A New Buffalo Born in the Huleh Valley." *Ha'aretz*, June 25, 2007. https://www.haaretz.co.il/misc/article-print -page/1.1420274.

Ashkenazy, Tuvia. *The Bedouins in the Land of Israel.* New York: Ariel Books, 1957.

Assad, Talal. *Genealogies of Religion: Discipline and Reasons of Power in Christianity and Islam.* Baltimore: Johns Hopkins University Press, 1993.

Atzmon, Ya'acov. "A Visit to Deganya Alef." *Cattle and Dairy Economy* 81 (1966): 2–9.

Atzmon, Ya'acov. "With Retirement." *Meshek Habakar Vehahalav* (1977): 3–8.

"Automating Dairy Farms." Afimilk. http://www.afimilk.com.

Aviam, Yisha'ayahu. "The World Congress for Fertility and Sterility Increasing the Prestige of Science in Israel." *Ma'ariv*, May 29, 1968.

Avitzur, Shmuel. *Changes in Eretz-Israel Agriculture, 1875–1975.* Tel Aviv: Milo Ltd. and Avshalom Institute, 1977.

Azaryahu, Maoz, and Arnon Golan. "Zionist Homelandscapes (and Their Constitution) in Israeli Geography." *Social and Cultural Geography* 5, no. 3 (2004): 497–513.

Baldensperger, Jean. "Palestine: An Account of Bee-Keeping There by an Eye-Witness." *American Bee Journal* 24, no. 4 (1888): 59–60.

Baldensperger, Philip. "Bees in Palestine." *Bee World* 12 (1931): 34–36.

Baldensperger, Philip. "The Identification of Ain-Rimmon with Ain-Urtas (Artas)." *Palestine Exploration Fund Quarterly Statement* 43 (1912): 201–2011.

Baldensperger, Philip. "Marketing Honey on the Shores of the Mediterranean." *Bee World* 11 (1930): 133–136.

Baldensperger, Philip. *The Immovable East: Studies of the People and Customs of Palestine.* Boston: Small, Maynard and Company, 1913.

Baldensperger, Philip. "Women in the East." *Palestine Exploration Fund Quarterly Statement* (1900): 66–67.

Bar-Adon, Pesah [Aziz Afandi]. *Among the Herds of Sheep (From the Stories of a Shepherd).* Tel Aviv: Eli'asaf Publishers, 1942.

Bar-Adon, Pesah [Aziz Afandi]. *In Desert Tents: From the Notes of a Hebrew Shepherd among Bedouin Tribes.* Jerusalem: Kiryat-Sefer, 1981.

Bar-Adon, Pesah [Aziz Afandi]. "A Sheep." *Mozna'im: Journal for Literature, Criticism, and Art* 3 (November 3, 1933): 2–4.

Barak, On. *On Time: Technology and Temporality in Modern Egypt.* Berkeley: University of California Press, 2013.

Barakat, Nora. "Marginal Actors? The Role of Bedouin in the Ottoman Administration of Animals as Property in the District of Salt, 1870–1912." *Journal of the Economic and Social History of the Orient* 58, no. 1–2 (2015): 105–134.

Barton, Gregory A. "Environmentalism, Development and British Policy in the Middle East 1945–56." *Journal of Imperial and Commonwealth History* 38, no. 4 (2010): 619–639.

Bar-Yosef, Eitan. *The Holy Land in English Culture 1799–1917: Palestine and the Question of Orientalism.* Oxford: Oxford University Press, 2005.

Bar-Yosef, Eitan. *Villa in the Jungle: Africa in Israeli Culture.* Jerusalem: Hakibbutz Hameuchad Publishing House, 2013.

"Beekeeping." *Filastin*, March 18, 1922, 1.

"Beekeeping in Palestine." *Filastin*, July 27, 1933, 7.

"Bee Supervision Command." *Davar*, August 26, 1928.

Beinart, William, and Lotte Hughes. *Environment and Empire.* Oxford: Oxford University Press, 2007.

Beker, Dov. *In the Meadows.* Israeli Shepherds Association, 1972.

Beker, Dov. *Sheep and Goat Rearing.* Ein-Harod: Hebrew Shepherds Association, 1948.

Bell, T. "Preliminary Work on the Artificial Insemination of Live Stock in Palestine." *Empire Journal of Experimental Agriculture* 6 (1938): 188.

Ben-Arzi, Yossi. "Religious Ideology and Landscape Formations: The Case of the German Templars in Eretz-Israel." In *Ideology and the Landscape in Historical Perspective*, edited by Alan R. H. Baker and Gideon Biger, 83–106. Cambridge: Cambridge University Press, 2006.

Ben-Bassat, Yuval. *Petitioning the Sultan: Justice and Protest in Late Ottoman Palestine.* London: I. B. Tauris, 2013.

Ben-David, Orit. "Tiyul as an Act of Consecration of Space." In *Grasping Land: Space and Place in Contemporary Israeli Discourse and Experience*, edited by Eyal Ben-Ari and Yoram Bilu. Albany: SUNY Press, 1997.

Ben-Naeh, Yaron. "'Thousands Great Saints': Evliua Çelebi in Ottoman Palestine." *Quest: Issues in Contemporary Jewish History* 6 (2013): 1–18.

Benninghaus, Christina. "Great Expectations: German Debates about Artificial Insemination in Humans around 1912." *Studies in History and Philosophy of Biological and Biomedical Sciences* 38 (2007): 374–392.

Benton, Frank. "A Bee-Convention in Syria." *American Bee Journal* 21, no. 35 (1885): 551.

Benton, Frank. "The New Races of Bees." *American Bee Journal* 20, no. 3 (1884): 38–39.

Berlovitch, Yafa. *Wounded Bird: Dora Bader—A Diary (1933–1937)*. Kibbutz Mizra' Hakibbutz Hame'uchad Publishers, 2011.

Berns, Andrew D. "The Place of Paradise in Renaissance Jewish Thought." *Journal of the History of Ideas* 75, no. 3 (2014): 351–371.

Besky, Sarah, and Alex Blanchette, eds. *How Nature Works: Rethinking Labor on a Troubled Planet*. Albuquerque: University of New Mexico Press, 2019.

Biger, Gideon, and Nili Lifshitz. "The Afforestation Policy of the British Government in the Land of Israel." *Ofakim Begeographia* 40–41 (1994): 5–16.

Bijker, Wiebe E. "Dikes and Dams, Thick with Politics." *Isis* 98, no. 1 (2007): 109–123.

Birenboim-Carmeli, Daphna. "Contested Surrogacy and the Gender Order in Israel." In *Assisting Reproduction, Testing Genes: Global Encounters with New Biotechnologies*, edited by Daphna Birenboim-Carmeli and Marcia Inhorn, 192–193. New York: Berghahn Books, 2009.

Birenboim-Carmeli, Daphna, and Yoram Carmeli, eds. *Kin, Gene, Community: Reproductive Technologies among Jewish Israelis*. New York: Berghahn Books, 2010.

Blanchette, Alex. *Porkopolis: American Animality, Standardized Life, and the Factory Farm*. Durham, NC: Duke University Press, 2020.

Blum, Israel Robert. "Beekeeping and Its Needs." *Davar*, December 14, 1932.

Blum, Israel Robert. *The Man and the Bee*. Tel Aviv: Tversky, 1951.

Blum, Israel Robert. "World Congress for Beekeepers." *Davar*, September 25, 1956, 4.

Bodenheimer, Fritz Simon. *Animal and Man in Bible Lands*. Leiden: E. J. Brill, 1960.

Bodenheimer, Fritz Simon. "Studies in Animal Populations II: Seasonal Population—Trends of Honey-Bee." *Quarterly Review of Biology* 12, no. 4 (1937): 406–425.

Boime, Albert. "William Holman Hunt's 'The Scapegoat': Rite of Forgiveness / Transference of Blame." *Art Bulletin* 84, no. 1 (2002): 94–114.

Bondi, Ruth. "Reception—For Animal Only." *Davar*, February 3, 1956, 20.

Bourdieu, Pierre. *Distinction: A Social Critique of the Judgment of Taste*. Cambridge, MA: Harvard University Press, 1984.

Bourdieu, Pierre. *Outline of a Theory of Practice*. Translated by R. Nice. Cambridge: Cambridge University Press, 1977.

Bower, Bruce. "Excavators Find Honey of a Discovery: Israeli Site Yields Oldest Known Example of Beekeeping." *Science News*, September 27, 2008, 11.

Braverman, Irus. *Planted Flags: Trees, Land, and Law in Israel/Palestine*. Cambridge: Cambridge University Press, 2009.

Braverman, Irus. "Planting the Promised Landscape: Zionism, Nature, and Resistance in Israel/Palestine." *Natural Resources Journal* 49 (2009): 317–361.

Braverman, Irus. "'The Tree Is the Enemy Soldier': A Sociolegal Making of War Landscapes in the Occupied West Bank." *Law and Society Review* 42, no. 3 (2008): 449–482.

Brentjes, Burchard. "Water Buffalo in the Cultures of the Ancient Near-East," *Zeirschrift für Saugetierkunde* 34 (1969): 187–191.

Bresalier, Michael. "From Healthy Cows to Healthy Humans: Integrated Approaches to World Hunger, c. 1930–1965." In *Animals and the Shaping of Modern Medicine: One Health and Its Histories*, edited by Abigail Woods, Michael Bresalier, Angela Cassidy, and Rachel Mason Dentinger, 119–160. Cham, Switzerland: Palgrave Macmillan, 2018.

Browne, Janet. *The Secular Ark: Studies in the History of Biogeography*. New Haven, CT: Yale University Press, 1983.

Buheiry, Marwan R. "The Agricultural Exports of Southern Palestine, 1885–1914." *Journal of Palestine Studies* 10, no. 4 (1981): 61–81.

Burke, Edmund III. "Pastoralism and the Mediterranean Environment." *International Journal of Middle East Studies* (2010): 663–665.

Butcher, Tim. "Israel No Longer Land of Milk and Honey after 60% Fall in Honey Harvest." *Telegraph*, September 16, 2008. http://www.telegraph.co.uk/earth/earthnews/3351841/Israel-no-longer-Land-of-Milk-and-Honey-after-60-fall-in-honey-harvest.html.

Callon, Michel. "Some Elements of a Sociology of Translation: Domestication of the Scallops and the Fishermen of St Brieuc Bay." In *Power, Action and Belief: A New Sociology of Knowledge?*, edited by John Law, 196–233. London: Routledge, 1986.

Canaan, Tawfiq. *Mohammedan Saints and Sanctuaries in Palestine*. London: Luzac and Co., 1927. Reprinted from *Journal of the Palestine Oriental Society*.

"A Car for Every One Hundred Thousand People in the Land of Israel." *Doar Hayom*, February 25, 1935, 1.

Carmel, Alex. "C. F. Spittler and the Activities of the Pilgrims Mission in Jerusalem." In *Ottoman Palestine 1800–1914: Studies in Economic and Social History*, edited by Gad G. Gilber, 255–286. Leiden: E. J. Brill, 1990.

Carmel, Alex. *The German Settlement in the Land of Israel in the Late Ottoman Period: National, Local, and International Problems*. Haifa: Haifa University Press, 1970.

Casto, E. Ray. "Economic Geography of Palestine." *Economic Geography* 13, no. 3 (1937): 235–259.

Chagall, Marc. "I and the Village." http://www.moma.org/collection/object.php ?object_id=78984.

Chatwin, Bruce. *The Songlines*. New York: Penguin Books, 1987.

Clarke, Adele E. *Disciplining Reproduction: Modernity, American Life and the Problem of Sex*. Berkeley: University of California Press, 1998.

Clarke, Adele E. "Reflections on the Reproductive Sciences in Agriculture in the UK and US, ca. 1900–2000+." *Studies in History and Philosophy of Biological and Biomedical Sciences* 38 (2007): 316–339.

Clarke, Adele E. "Research Materials and Reproductive Science in the United States, 1910–1940." In *Physiology in the American Context 1850–1940*, edited by Gerald L. Geison, 323–350. New York: Springer, 1987.

Cohen, Amnon. *Economic Life in Ottoman Jerusalem*. Cambridge: Cambridge University Press, 1989.

Cohen, Amnon. "Ottoman Rule and the Re-Emergence of the Coast of Palestine (17th–18th Centuries)." *Revue de l'Occident musulman et de la Méditerranée* 39 (1985): 163–175.

Cohen, Amnon, and Bernard Lewis. *Population and Revenue in the Towns of Palestine in the Sixteenth Century*. Princeton, NJ: Princeton University Press, 1978.

Cohen, Hillel. *Army of Shadows: Palestinian Collaboration with Zionism, 1917–1948*. Translated by Haim Watzman. Berkeley: University of California Press, 2008.

Cohen, Hillel. *1929: Year Zero of the Jewish-Arab Conflict*. Jerusalem: Keter, 2013.

Cohen, Shaul E. *The Politics of Planting: Israeli-Palestinian Competition for Control of Land in the Jerusalem Periphery*. Chicago: University of Chicago Press, 1993.

Comaroff, Jean, and John L. Comaroff. *Of Revelation and Revolution, Volume 1: Christianity, Colonialism, and Consciousness in South Africa*. Chicago: University of Chicago Press, 1991.

Cooper, Alix. *Inventing the Indigenous: Local Knowledge and Natural History in Early Modern Europe*. New York: Cambridge University Press, 2007.

"Course for Beekeepers." *Davar*, January 10, 1933, 4.

"The Cow That Won Two Rewards—Zkufa." *Davar*, December 24, 1937, 3.

"Cows from Palestine for Improving the Cattle Breeds in Iraq and Ceylon." *Haqiqat al-Amr*, December 16, 1941, 3.

Cox-Foster, Diana L., Sean Conlan, Edward C. Holmes, Gustavo Palacios, Jay D. Evans, Nancy A. Moran, Phenix-Lan Quan, et al. "A Metagenomic Survey of Microbes in Honey Bee Colony Collapse Disorder." *Science* 318 (2007): 283–287.

Crane, Eva. *The World History of Beekeeping and Honey Hunting*. New York: Routledge, Chapman, and Hall, 1999.

Cronon, William. *Changes in the Land: Indians, Colonists, and the Ecology of New England*. New York: Hill and Wang, 1983.

Cronon, William. *Nature's Metropolis: Chicago and the Great West*. New York: W. W. Norton and Company, 1991.

Crosby, Alfred W. *Ecological Imperialism: The Biological Expansion of Europe, 900–1900*. Cambridge: Cambridge University Press, 1986.

Crowfoot, Grace M., and Louise Baldensperger. *From Cedar to Hyssop: A Study in the Folklore of Plants in Palestine*. London: Sheldon Press, 1932.

"Dairy Cattle Exhibition in Emek Izrael." *Cowboy Pages* 1 (1958): 1.

Daston, Lorraine, and Gregg Mitman, eds. *Thinking with Animals: New Perspectives on Anthropomorphism*. New York: Columbia University Press, 2005.

Davis, Diana K. *Resurrecting the Granary of Rome: Environmental History and French Colonialism Expansion in North Africa*. Athens: Ohio University Press, 2007.

Davis, Diana K., and Edmund Burke III, eds. *Environmental Imaginaries of the Middle East and North Africa*. Athens: Ohio University Press, 2011.

Davis, Natali Zemon. *Fiction in the Archives: Pardon Tales and Their Tellers in Sixteenth-Century France*. Stanford, CA: Stanford University Press, 1987.

Degani, Arnon. "An Invitation to Expand Zionism's Defining Contours." *Hazman Haze*, March 2019, https://hazmanhazeh.org.il/degani.

Department of Agriculture and Forests. "Sale of Sugar to Beekeepers." *Palestine Post*, October 7, 1934.

Divan, Andrew H. "First Queen by Mail from Jerusalem." *American Bee Journal* 20, no. 51 (1884): 809.

Dori, Shlomo. *Meshek Habakar Vehahalav* (1977): 9.

Dori, Shlomo. *News from the Past: Chapters in the History of Dairy Cattle Farming in Israel, Part 1*. Caesarea: Cattle Breeders Association in Israel, 1992.

Dori, Shlomo. *News from the Past: Chapters in the History of Dairy Cattle Farming in Israel, Part 2*. Caesarea: Cattle Breeders Association in Israel, 1996.

Dori, Shlomo. *News from the Past: Chapters in the History of Dairy Cattle Farming in Israel, Part 3*. Caesarea: Cattle Breeders Association, 2002.

Douglas, Mary. *Purity and Danger: An Analysis of Concepts of Pollution and Taboo*. New York: Frederich A. Praeger Publishers, 1966.

Doumani, Beshara. *Rediscovering Palestine: Merchants and Peasants in Jabal Nablus, 1700–1900*. Berkeley: University of California Press, 1995.

Doumani, Beshara. "Review of Thomas Philipp, Acre: The Rise and Fall of a Palestinian City, 1730–1831." *Journal of Palestine Studies* 33, no. 1 (2003): 98–100.

Drayton, Richard. *Nature's Government: Science, Imperial Britain and the "Improvement" of the World*. New Haven, CT: Yale University Press, 2000.

Dror, Yaron. "Udderly Marvelous Gina: Israel's Most Productive Cow." *Ha'aretz*, June 2, 2004, 8.

DuPuis, Melanie. *Nature's Perfect Food: How Milk Became America's Drink*. New York: NYU Press, 2002.

Ecole Nationale d'Ingeniéur, Sud Alsace, Mulhouse. "Graduates." http://www.anciens -ensisa.org/.

Edgerton, David. *The Shock of the Old: Technology and Global History Since 1900*. Oxford; New York: Oxford University Press, 2007.

Edwards, Paul N., and Gabrielle Hecht. "History and the Technopolitics of Identity: The Case of Apartheid South Africa." *Journal of Southern African Studies* 3 (2010): 619–639.

"8937k of Milk—The Annual Yield of One Cow." *Davar*, November 16, 1937, 6.

Eilat [Esptein], Eliyahu. *The Bedouins: Their Lives and Customs*. Tel Aviv: A. Y. Stibel, 1933.

Eilenbogen, W. "Infertility in Cattle: Reasons and Ways of Treating." *Hasade* 13 (1933): 12–15.

Elazari-Volcani, Yitzhak. *The Dairy Industry as the Basis for Colonisation in Palestine*. Tel Aviv: Palestine Economic Society, 1928.

Elazari-Volcani, Yitzhak. *The Design of the Agriculture of the Land (of Israel)*. Rehovot Experimental Station: Jewish Agency, 1937.

Elazari-Volcani, Isaac. *The Fellah's Farm*. Tel Aviv: Jewish Agency for Palestine, Institute of Agriculture and Natural History, Agricultural Experiment Station, 1930.

Elazari-Volcani, Yitzhak. *On the Road*. Jaffa: Hapo'el Hatza'ir, 1918.

Elazari-Volcani, Yitzhak. "On the State of Farming." *Hapoel Hatzair*, February 11, 1912, 7–9.

El-Eini, Roza. "British Forestry Policy in Mandate Palestine, 1929–48." *Middle Eastern Studies* 35, no. 3 (1999): 72–155.

El-Eini, Roza. "Governmental Fiscal Policy in Mandatory Palestine in the 1930s." *Middle Eastern Studies* 33, no. 3 (1997): 570–596.

El-Eini, Roza. "The Implementation of British Agricultural Policy in Palestine in the 1930s." *Middle Eastern Studies* 32, no. 4 (1996): 211–250.

El-Eini, Roza. *Mandated Landscape: British Imperial Rule in Palestine, 1929–1948*. New York: Routledge, 2006.

Elmaliach, Tal. *The Kibbutz Industry, 1923–2007: Discussing Questions of Economy and Society*. Givat Haviva: Yad Ya'ari Publishers, 2009.

Erlanger, Steven. "After 2,000 Years, a Seed from Ancient Judea Sprouts." *New York Times*, June 12, 2005.

Eshel, Ruth. "Dancing on the Sidewalks of Ein-Hashofet." *Dance Now* 14, no. 1 (2000): 44–48.

Essaid, Aida A. *Zionism and Land Tenure in Mandate Palestine*. New York: Routledge, 2014.

Etkes, Dror. *Israeli Settler Agriculture as a Means of Land Takeover in the West Bank*. Jerusalem: Kerem Navot, 2013.

Etsion, Lotek. Interview by Tamar Novick, Kibbutz Merhavia, October 18, 2011.

Ever-Hadani, Aaron. *Agriculture and Settlement in Israel: A Decade after Its Establishment*. Beit Dagan: Ministry of Agriculture, 1958.

"Expediting the Fellahs." *Davar*, December 14, 1932.

Eyal, Gil. *The Disenchantment of the Orient: Expertise in Arab Affairs and the Israeli State*. Stanford, CA: Stanford University Press, 2006.

Fabian, Johannes. *Time and the Other: How Anthropology Makes Its Object*. New York: Columbia University Press, 1983.

Fairhead, James, and Melissa Leach. *Misreading the African Landscape: Society and Ecology in a Forest-Savana Mosaic*. Cambridge: Cambridge University Press, 1996.

Falah, Ghazi. "How Israel Controls the Bedouin in Israel." *Journal of Palestine Studies* 14, no. 2 (1985): 35–51.

Falk, Rafael. *Zionism and the Biology of the Jews*. Tel Aviv: Resling, 2006.

"Famous for Honey: An Industry of Palestine in Biblical Days May Be Revived." *Kaskell Free Press*, July 28, 1900.

"Food and Agricultural Commodity Production." Food and Agriculture Organization of the United Nations. http://faostat.fao.org/site/339/default.aspx.

Finkelstein, Michael. "Professor Bernhard Zondek—An Interview." *Journal of Reproduction and Fertility* 12 (1966): 3–19.

Firestone, Ya'akov. "Crop-Sharing Economics in Mandatory Palestine—Part I." *Middle Eastern Studies* 11, no. 1 (1974): 3–23.

Fitzgerald, Deborah. *Every Farm a Factory: The Industrial Ideal in American Agriculture*. New Haven, CT: Yale University Press, 2003.

Fogel-Bijawi, Sylvia. "Families in Israel: Postmodernity, Feminism and the State." *Journal of Israeli History* 21, no. 2 (2002): 38–62.

Folman, Yeshayahu, Amiel Berman, Zeev Herz, Moshe Kaim, Miriam Rosenberg, Meir Mamen, and Sali Gordin. "Milk Yield and Fertility of High-Yielding Dairy Cows in a Sub-Tropical Climate during Summer and Winter." *Journal of Dairy Research* 46, no. 3 (1979): 411–425.

Forman, Geremy, and Alexander Kedar. "Colonialism, Colonization and Land Law in Mandate Palestine: The Zor al-Zarqa and Barrat Qisarya Land Disputes in Historical Perspective." *Theoretical Inquiries in Law* 4, no. 2 (2003): 491–540.

Foucault, Michel. *The History of Sexuality, Volume I: An Introduction.* London: Allen Land, 1979.

Franklin, Sarah. *Dolly Mixtures: The Making of Genealogy.* Durham, NC: Duke University Press, 2007.

Franzman, Seth J., and Ruth Kark. "Bedouin Settlement in Late Ottoman and British Mandatory Palestine: Influence on the Cultural and Environmental Landscape, 1870–1948." *New Middle Eastern Studies* 1 (2011): 1–23.

Freund, Dr. S., and Dr. A. Rosen. *The Sterility of Cows.* Tel Aviv: Agricultural Experimental Station, Zionist Organization, 1925.

Fudge, Erica. *Quick Cattle and Dying Wishes: People and Their Animals in Early Modern England.* Ithaca, NY: Cornell University Press, 2018.

Gaard, Greta. "Toward a Feminist Postcolonial Milk Studies." *American Quarterly* 65, no. 3 (2013): 595–618.

Gaudillière, Jean-Paul. "Better Prepared than Synthesized: Adolf Butenandt, Schering AG, and the Transformation of Sex Steroids into Drugs (1930–1946)." *Studies in History and Philosophy of Biological and Biomedical Sciences* 36, no. 4 (2005): 612–644.

Gerber, Haim. "Modernization in Nineteenth-Century Palestine: The Role of Foreign Trade." *Middle Eastern Studies* 18, no. 3 (1982): 250–264.

Gere, Cathy. *Knossos and the Prophets of Modernism.* Chicago: University of Chicago Press, 2009.

Ghanim, Honaida. "What Is the Color of the Arab? A Critical Look at Color Play." In *Racism in Israel*, edited by Yehuda Shenhav and Yossi Yona, 76–92. Tel Aviv: Hakibbutz Hameuchad Publishing House, 2008.

Gil, Moshe. "The Decline of the Agrarian Economy in Palestine under Roman Rule." *Journal of the Economic and Social History of the Orient* 49, no. 3 (2006): 285–328.

Gilad, Efrat. "'The Child Needs Milk and Milk Needs a Market': The Politics of Nutrition in the Interwar Yishuv." *Gastronomica: The Journal for Food Studies* 21, no.1 (2021): 7–16.

Gilad, Efrat. "Meat in the Heat: A History of Tel Aviv under the British Mandate for Palestine (1920s–1940s)." PhD diss., University of Geneva, 2021.

Gilbert, S. J. "An Unusual Strain of Brucella Causing Abortion of Cattle in Palestine." *Journal of Comparative Pathology and Therapeutics* 43 (1930): 118–124.

Gimlin, Debra. "What Is 'Body Work'? A Review of the Literature." *Sociology Compass* 1, no. 1 (2007): 353–370.

Ginsberg, Ralph. "How Did the Israeli Holstein Cow Become a World Leader in Milk Yields?" Israel Dairy Board. https://www.israeldairy.com/israeli-holstein-cow-become -world-leader-milk-yields-3/.

Gitlin, Marcia. "Bernhard Zondek—A Great Doctor in Israel." *Zionist Record*, September 21, 1951.

Golan, Tal. "Introduction: Special Issue—Science, Technology, and Israeli Society." *Israel Studies* 9, no. 2 (2004): iv–viii.

Goldstein, Yaakov. *The Shepherds Fraternity*. Tel Aviv: Ministry of Defense Publications, 1993.

Goodrich-Fereer, Ada. *Arabs in Tent and Town: An Intimate Account of the Family Life of the Arabs of Syria, Their Manner of Living in Desert and Town, Their Hospitality, Customs and Mental Attitude, with a Description of the Animals Birds, Flowers and Plants of Their Country*. New York: G. P. Putnam's Sons, 1924.

Goren, Haim. *Go Research the Land: German Studies of the Land of Israel in the Nineteenth Century*. Jerusalem: Yad Ben-Zvi, 1999.

Goren, Haim. *True Catholics and Good Germans: The Catholic Germans and the Land of Israel, 1838–1910*. Jerusalem: Hebrew University Magnes Press, 2004.

Goren, Haim, and Richab Rubin. "This Is How Modern Beekeeping Started in This Country." *Mada* 29, no. 4–5 (1985). https://web.archive.org/web/20140907225240 /http://www1.snunit.k12.il//heb_journals/mada/294181.html.

Granqvist, Hilma. *Birth and Childhood among the Arabs: Studies in a Huhnmadan Village in Palestine*. Helsinki: Söderström ja Co Förlagsaktiebolag, 1947.

Granqvist, Hilma. *Marriage Conditions in a Palestinian Village*. Helsinki: Centraltryckeri och Bokbinderi, 1931.

Greenberg, Zalman, and Anton Alexander. "Israel Jacob Kligler: The Story of 'A Little Big Man.'" *Korot—The Israeli Journal of the History of Medicine and Science* 21 (2012): 175–206.

Greenwood, Anthony. "Istanbul's Meat Provisioning: A Story of the Celepkeşan System." PhD diss., University of Chicago, 1988.

Grove, Richard H. *Green Imperialism: Colonial Expansion, Tropical Island Edens and the Origins of Environmentalism, 1600–1860*. Cambridge: Cambridge University Press, 1995.

Gutman, Mordechai. "How Have We Coped with Infertility in Our Herd?" *Hasade* 20, no. 9 (1940): 457–458.

Ha'am, Ahad. "Truth from the Land of Israel." *Hamelitz*, June 30, 1891.

Halevi, Eli'ezer. *A Journey to the Land of Israel*. Tel Aviv: Omanut Publishers, 1931. First published 1838.

Halperin, Haim. *Changes in Jewish Agriculture*. Tel Aviv: Ayanot, 1954.

Hamdy, Sherine. *Our Bodies Belong to God: Organ Transplants, Islam, and the Struggle for Human Dignity in Egypt*. Berkeley: University of California Press, 2012.

Haraway, Donna. *Primate Visions: Gender, Race, and Nature in the World of Modern Science*. New York: Routledge, 1989.

Hashiloni-Dolev, Yael. *The Fertility Revolution*. Ben-Shemen: Modan, 2013.

Hasson, Nir. "In an Attempt to Taste the Ancient Fruit, Researchers Germinated Six 2,000-Year-Old Date Seeds." *Ha'aretz*, February 6, 2020. https://www.haaretz.co.il /science/archeology/1.8501814.

Headrick, Daniel. *Power over People: Technology, Environment, and Imperialism*. Princeton, NJ: Princeton University Press, 2009.

Hecht, Gabrielle. "Rupture-Talk in the Nuclear Age: Conjuring Colonial Power in Africa." *Social Studies of Science* 32, no. 5–6 (2002): 691–727.

Hirsch, Dafna. "'Hummus Is Best When It Is Fresh and Made by Arabs': The Gourmetization of Hummus in Israel and the Return of the Repressed Arab." *American Ethnologist* 38 (2011): 617–630.

Hirsch, Dafna, and Ofra Tene. "Hummus: The Making of an Israeli Culinary Cult." *Journal of Consumer Culture* 13, no. 1 (2013): 25–45.

Hirsch, Dr. Siegfried. "Sheep and Goats in Palestine." *Bulletin of the Palestine Economic Society* 6, no. 2 (1933): 1–77.

Hodge, Joseph Morgan. *Triumph of the Expert: Agrarian Doctrines of Development and the Legacies of British Colonialism*. Athens: Ohio University Press, 2007.

"Honey in Jerusalem." *Bruce Herald*, February 1, 1901.

Hribal, Jason. *Fear of the Animal Planet: The Hidden History of Animal Resistance*. Chico, CA: AK Press, 2010.

Hughes, Matthew. "From Law and Order to Pacification: Britain's Suppression of the Arab Revolt in Palestine, 1936–39." *Journal of Palestine Studies* 39, no. 2 (2010): 6–22.

Hughes, Thomas P. *Networks of Power: Electrification in Western Society, 1880–1930*. Baltimore: Johns Hopkins University Press, 1993.

Hütteroth, Wolf-Dieter, and Kamal Abdulfattah. *Historical Geography of Palestine, Transjordan and Southern Syria in the Late 16th Century*. Erlangen, Germany: Franconian Geographic Society, 1977.

Ibn Al-Sheikh. "Milk—Important Information for Farmers and Non-Farmers." *Al-Liwaa*, February 18, 1936, 7.

Ilani, Ofri. "2,000-Year-Old Date Seed Grows in the Arava." *Ha'aretz*, February 15, 2007. http://www.haaretz.com/print-edition/news/2-000-year-old-date-seed-grows-in -the-arava-1.213054.

"In the Holy Land." *Hamelitz*, July 5, 1900.

Inhorn, Marcia. *Infertility and Patriarchy: The Cultural Politics of Gender and Family Life in Egypt*. Philadelphia: University of Pennsylvania, 1996.

İslamoğlu, Huri. "Property as a Contested Domain: A Reevaluation of the Ottoman Land Code of 1858." In *New Perspectives on Property and Land in the Middle East*, edited by Roger Owen, 3–61. Cambridge, MA: Harvard University Press, 2000.

İslamoğlu, Huri. *State and Peasant in the Ottoman Empire: Agrarian Power Relation and Regional Economic Development in Ottoman Anatolia during the Sixteenth Century*. Leiden: Brill, 1994.

Israeli Honey Council. *On the History of Beekeeping in Israel: Problems and Related Stories*. http://www.honey.org.il/info/about-us/begin_peer.htm.

Jacks, G. V., and R. O. Whyte. *The Rape of the Earth: A World Survey of Soil Erosion*. London: Faber and Faber Ltd., 1939.

Jacoby, Karl. *Crimes against Nature: Squatters, Poachers, Thieves, and the Hidden History of American Conservation*. Berkeley: University of California Press, 2001.

Jaffa, the Orange's Clockwork. Directed by Eyal Sivan. 2009.

Jarvis, Major C. S. "The Arab, the Goat, and the Camel: Destroyers of the Desert." *Palestine Post*, October 11, 1934.

"Jewish Beekeepers Conference." *Davar*, August 30, 1939.

Jonas, Salo. "Cattle Raising in Palestine." *Agricultural History* 26, no. 3 (1952): 93–104.

Jones, Tobi Craig. *Desert Kingdom: How Oil and Water Forded Modern Saudi Arabia*. Cambridge, MA: Harvard University Press, 2010.

Kababia, Dorit. Interview by Tamar Novick, Beit Dagan, September 15, 2013.

Kabha, Mustafa, and Nahum Karlinsky. *The Lost Orchard: The Palestinian-Arab Citrus Industry, 1850–1950*. Syracuse, NY: Syracuse University Press, 2021.

Kahn, Dorothy. "Flowing with Honey: How It's Done in a Jewish Settlement." *Palestine Post*, August 30, 1938, 6.

Kahn, Susan Martha. *Reproducing Jews: A Cultural Account of Assisted Conception in Israel*. Durham, NC: Duke University Press, 2000.

Kanaaneh, Rhoda Ann. *Birthing the Nation: Strategies of Palestinian Women in Israel*. Berkeley: University of California Press, 2002.

Kanafani, Ghassan. "The Land of Sad Oranges." In *Men in the Sun and Other Palestinian Stories*, translated by Hilary Kilpatrick, 75–81. Boulder, CO: Lynne Rienner Publishers, 1999.

Kardeshian, Kirk. *Milk Money: Cash, Cows, and the Death of the American Dairy Farm*. Lebanon: University of New Hampshire Press, 2012.

Kark, Ruth. "Millenarism and Agricultural Settlement in the Holy Land in the Nineteenth Century." In *Deutsche in Palästina und ihr Anteil an der Modernisierung des Landes*, edited by Jacob Eisler, 14–29. Wiesbaden: Harrassowitz Verlag, 2008.

Karlinsky, Nahum. *California Dreaming: Ideology, Society, and Technology in the Citrus Industry of Palestine 1890–1939*. Albany: SUNY Press, 2005.

Karlinsky, Nahum, and Mustafa Kabha. "The Missing Orchard: Palestinian-Arab Citrus Cultivation before 1948." *Zmanim: A Historical Quarterly* 129 (2015): 94–109.

Karmon, Yehuda. "The Drainage of the Huleh Swamps." *Geographical Review* 50, no. 2 (1960): 169–193.

Katz, Saul. "'The First Furrow': Ideology, Settlement, and Agriculture in Petah-Tikva in Its First Decade." *Catedra* 23 (1982): 124–157.

Katz, Saul. "Sociological Aspects in the Development (and Replacement) of Agricultural Knowledge in Israel: The Emergence of Ex-Scientific Systems for the Production of Agricultural Knowledge, 1880–1940." PhD diss., Hebrew University of Jerusalem, 1986.

Keith-Roach, Edward. "Changing Palestine." *National Geographic Magazine*, April 1934, 493–527.

Kelner, Shaul. "Ethnographers and History." *American Jewish History* 98, no. 1 (2014): 17–22.

Khalidi, Rashid. *The Iron Cage: The Story of the Palestinian Struggle for Statehood*. Boston: Beacon Press, 2006.

Khawalde, Sliman, and Dan Rabinowitz. "Race from the Bottom of the Tribe That Never Was: Segmentary Narratives amongst the Ghawarna of Galilee." *Journal of Anthropological Research* 58, no. 2 (2002): 225–243.

Khoury, Elias. *Bab al-Shams*. Beirut: Dar el Adab, 1998.

Kimmerling, Baruch, and Joel S. Migdal. *Palestinians: The Making of People*. Cambridge, MA: Harvard University Press, 1994.

Kohler, Robert E. "Lab History: Reflections," *Isis* 99 (2008): 761–768.

Kohler, Robert E. *Lords of the Fly*. Chicago: University of Chicago Press, 1994.

"Land of Promise." Spielberg Jewish Film Archive. http://www.youtube.com/watch?v=QDoD6W2z01s.

Langstroth, L. L. *The Hive and the Honeybee: The Classic Beekeeper's Manual*. Mineola, NY: Dover Publications, 1878.

Lansing, J. Stephen. *Priests and Programmers: Technologies of Power in the Engineered Landscape of Bali*. Princeton, NJ: Princeton University Press, 1991.

Larsson, Teodore. "A Visit to the Mat Makers of Huleh." *Palestine Exploration Fund Quarterly Statement* 68, no. 4 (1936): 225–230.

Latour, Bruno. "Give Me a Laboratory and I Will Raise the World." In *The Science Studies Reader*, edited by Mario Biagioli, 258–275. New York: Routledge, 1999.

Latour, Bruno. *We Have Never Been Modern*. Cambridge, MA: Harvard University Press, 1993.

Lavi, Zvi. "Miriam Baratz Learned How to Milk with the Bedoiuns." *Ma'ariv*, October 10, 1960, 10.

Leibler, Anat. "Statisticians' Ambition: Governmentality, Modernity and National Legibility." *Israel Studies* 9, no. 2 (2004): 121–149.

Levi, Isar. "'The Troubadour' of Sheep Management." In *David Zamir (A Collection for His Memory)*. Merhavia: Shepherds Association and Kibbutz Merhavia, 1968.

Levi, Uriel. *The History of Dairy Bovine in Israel*. Tel Aviv: Israeli Breeders Association, 1983.

Likhovski, Assaf. "Between 'Mandate' and 'State': Re-Thinking the Periodization of Israeli Legal History." *Journal of Israeli History* 19, no. 2 (1998): 39–68.

Livneh, Neri. "My Million Children—An Interview with Bruno Lunenfeld." *Ha'aretz*, May 29, 2002.

"Loans to Beekeepers." *Davar*, October 24, 1935.

Lockman, Zachary. *Comrades and Enemies: Arab and Jewish Workers in Palestine, 1906–1948*. Berkeley: University of California Press, 1996.

Lorimer, Hayden. "Walking: New Forms and Spaces for the Study of Pedestrianism." In *Geographies of Mobilities: Practices, Spaces, Subjects*, edited by Tim Cresswell, 19–34. Burlington, VT: Ashgate Publishing, 2011.

Lubani, Muntaha, and Ibtisam Abu Salim. "Interview with Muhammad Qasim Muhammad." Palestinian Oral History Archive, American University of Beirut Libraries, October 30, 1998. https://libraries.aub.edu.lb/poha/Record/4695.

Luke, Harry C., and Edward Keith-Roach. *The Handbook of Palestine, Issued under the Authority of the Government of Palestine*. London: Macmillian and Co., 1922.

Lunenfeld, Bruno. "Historical Perspectives in Gonadotrophin Therapy." *Human Reproduction Update* 10, no. 6 (2004): 453–467.

Lunenfeld, Bruno. Interview by Tamar Novick, Tel Aviv, August 30, 2013.

Lunenfeld, Bruno. "Management of Infertility: Past, Present and Future (from a Personal Perspective)." *Reproductive Medicine and Endocrinology* 10 (2013): 11–20.

"Machinery, Glass of Milk a Day, and Agricultural Experiments." *Al-Difa'*, April 9, 1943, 2.

Mahmoud, Rakan. "Interview with Ahmad Ismail Dakhloul." Palestine Remembered, January 18, 2010. https://www.palestineremembered.com/Safad/al-Salihiyya /Story17996.html.

Mahmoud, Rakan. "Interview with Warda al-Abdallah." Palestine Remembered, March 10, 2010. https://www.palestineremembered.com/Safad/Mallaha/Story17992 .html.

Malchi, Yossi. "Modernity, Nationality, and Society: The Beginning of Hebrew Flight in Palestine 1932–1940." MA thesis, Tel Aviv University, 2007.

Marchenay, Phillipe. *L'Homme et l'Abeille*. Paris: Berger-Levrault, 1979.

Marks, Lara. *Sexual Chemistry: A History of the Contraceptive Pill*. New Haven, CT: Yale University Press, 2001.

Marx, Leo. *The Machine in the Garden: Technology and the Pastoral Ideal in America*. Oxford: Oxford University Press, 1964.

Massad, Joseph A. *The Persistence of the Palestinian Question: Essay on Zionism and the Palestinians*. London: Routledge, 2006.

Mauss, Marcel. "Techniques of the Body." *Economy and Society* 2, no. 1 (1973 [1936]): 70–88.

Maxximilk—Much More Milk. http://www.maxximilk.com/index.php?tab=home.

Mazar, Amihai, and Nava Panitz-Cohen. "It Is the Land of Honey: Beekeeping at Tel Reḥov." *Near Eastern Archaeology* 70, no. 4 (2007): 202–219.

McCann, James C. *Green Land, Brown Land, Black Land: An Environmental History of Africa, 1800–1990*. Portsmouth. NH: Heinemann, 1999.

Meiton, Fredrik. *Electrical Palestine: Capitalism and Technology from Empire to Nation*. Oakland: University of California Press, 2019.

Melville, Elinor G. K. *A Plague of Sheep: Environmental Consequences of the Conquest of Mexico*. New York: Cambridge University Press, 2005.

Mennell, Philip. *The Coming Colony: Practical Notes on Western Australia*. London: Huntington and Co., 1892.

Merrill, Saleh. "Honey Producing in Old Palestine." *American Bee Journal* 40, no. 23 (1900): 356.

Merton, Robert K. "Science, Technology and Society in Seventeenth Century England." *Osiris* 4 (1938): 360–632.

Metzer, Jacob. *The Divided Economy of Mandatory Palestine*. Cambridge: Cambridge University Press, 1998.

Mikhail, Alan. *The Animal in Ottoman Egypt*. Oxford: Oxford University Press, 2013.

"Milk Farce in the Midst of Tragedy." *Filastin*, March 13, 1937, 1.

Mitchell, Timothy. *Carbon Democracy: Political Power in the Age of Oil*. London: Verso, 2011.

Mitchell, Timothy, ed. *Questions of Modernity*. Minneapolis: University of Minnesota Press, 2000.

Mitchell, Timothy. *Rule of Experts: Egypt, Techno-Politics, Modernity*. Berkeley: University of California Press, 2002.

Mitchell, W. J. T. "Holy Landscape: Israel, Palestine, and the American Wilderness." *Critical Inquiry* 26, no. 2 (2002): 193–223.

Monzote, Reinaldo Funes. "Ubra Blanca and the Politics of Milk in Socialist Cuba." *Oxford Research Encyclopedia for Latin American History*. Oxford: Oxford University Press, 2019.

Morag, M., A. A. Degen, and F. Popliker. "The Reproductive Performance of German Mutton Merino Ewes in a Hot Arid Climate." *Journal of Animal Breeding and Genetics* 89 (1972): 340–345.

"More than Two Hundred Cows to Iran—Israel Is the Biggest Cattle Supplier to That Country." *Davar*, December 4, 1967, 3.

Morris, Benny. *The Birth of the Palestinian Refugee Problem, 1947–1949*. Cambridge: Cambridge University Press, 1987.

Moscrop, John Jams. *Measuring Jerusalem: The Palestine Exploration Fund and British Interest in the Holy Land*. London: Leicester University Press, 2000.

Moshkovitz, Israel. "World Record for the Israeli Cow: 11,400 Litter Per Year." *Yedi'ot Achronot, Economy Supplement*, May 29, 2014, :10.

Mukerji, Chandra. *Impossible Engineering: Technology and Territoriality on the Canal du Midi*. Princeton, NJ: Princeton University Press, 2009.

Mundy, Martha, and Richard Saumarez Smith. *Governing Property, Making the Modern State: Law, Administration, and Production in Ottoman Syria*. London: I. B. Tauris, 2007.

N., F. S. "Land of the Promise at the Astor." *New York Times*, November 21, 1935.

Nadan, Amos. "Colonial Misunderstanding of an Efficient Peasant Institution: Land Settlement and Mushāʿ Tenure in Mandate Palestine, 1921–47." *Journal of Economic and Social History of the Orient* 46, no. 3 (2003): 320–354.

Nadan, Amos. *The Palestinian Peasant Economy under the Mandate: A Story of Colonial Bungling*. Cambridge, MA: Harvard University Press, 2006.

Naili, Falestin. "Henri Baldensperger: An Alsatian Missionary and 'Living Together' in Ottoman Palestine." *L'Annuaire de la Société d'Histoire de la Hardt et du Ried* 23 (2011): 1–22.

Neria, Ami. *Veterinary Medicine in the Land of Israel: 50 Years of Veterinarian Medicine, 1917–1967*. Tel Aviv: Re'emim Press, 2001.

Neumann, Boaz. *Land and Desire in Early Zionism*. Tel Aviv: Am Oved Publishers, 2009.

Noble, David F. *The Religion of Technology: The Divinity of Man and the Spirit of Invention*. New York: Penguin Books, 1999.

Nordlund, Christer. *Hormones of Life: Endocrinology, the Pharmaceutical Industry, and the Dream of a Remedy for Sterility, 1930–1970*. Sagamore Beach, MA: Watson Publishing International, 2011.

Norris, Jacob. *Land of Progress: Palestine in the Age of Colonial Development, 1905–1948*. Oxford: Oxford University Press, 2013.

Norton, David. *A History of the Bible as Literature, Part II: From 1700 to the Present Day*. Cambridge: Cambridge University Press, 1993.

Novick, Tamar. "All about Stavit: A Beastly Biography." *Teoria Vebikoret* 51 (2019): 15–40.

Novick, Tamar. "The Hand, the Hoe, and the Hose: Contested Technologies of Settler Colonialism." In *The Cambridge Handbook for the History of Technology, Vol. II*, edited by Dagmar Schäfer, Francesca Bray, Mattero Valleriani, Tiago Saravia, and Shadreck Chirikure, forthcoming.

Noy-Meir, Imanuel, and Talya Oron. "Effects of Grazing on Geophytes in Mediterranean Vegetation." *Journal of Vegetation Science* 12, no. 6 (2001): 749–760.

"Obituary Notices: Nora Baldensperger." *Bee Journal* 58 (1977): 128.

"Of Their Methods for Destroying the Resources of Arabs!" *Filastin*, January 29, 1947, 2.

Olszynko-Gryn, Jesse. "The Demand for Pregnancy Testing: The Aschheim-Zondek Reaction, Diagnostic Versatility, and Laboratory Services in 1930s Britain." *Studies in History and Philosophy of Biological and Biomedical Sciences* 47 (2014): 233–247.

Olszynko-Gryn, Jesse. *A Woman's Right to Know: Pregnancy Testing in Twentieth-Century Britain*. Cambridge, MA: MIT Press, 2023.

Orenstein, Daniel, Alon Tal, and Char Miller, eds. *Between Ruin and Restoration: An Environmental History of Israel*. Pittsburgh: University of Pittsburgh Press, 2013.

Orland, Barbara. "Turbo-Cows: Producing a Competitive Animal in the Nineteenth and Early Twentieth Centuries." In *Industrializing Organisms: Introducing Evolutionary History*, edited by Susan R. Schrepfer and Philip Scranton, 167–189. New York: Routledge, 2003.

Osborn, Henry S. *Palestine: Past and Present, with Biblical, Literary and Scientific Notices*. Philadelphia: James Challen and Son, 1859.

Oudshoorn, Nelly. *Beyond the Natural Body: An Archaeology of Sex Hormones*. New York: Routledge, 1994.

"Palestine." *Geographical Teacher* 11, no. 6 (1922): 359–366.

Palmon, Moshe. Interview by Nir Mann. Tel Aviv, May 27, 2001.

Palvitch, Israel. Interview by Nir Mann. Kibbut Ma'abarot, 1996.

Parreñas, Juno Salazar. *Decolonizing Extinction: The Work of Care in Orangutan Rehabilitation*. Durham, NC: Duke University Press, 2018.

Pawley, Emily. "Feeding Desire: Generative Environments, Meat Markets, and the Management of Sheep Intercourse in Great Britain, 1700–1750." *Osiris* 33, no. 1 (2018): 47–62.

Pawley, Emily. "The Point of Perfection: Cattle Portraiture, Bloodlines, and the Meaning of Breeding." *Journal of the Early Republic* 36, no. 1 (2016): 37–72.

Paxson, Heather. *The Life of Cheese: Crafting Food and Value in America*. Berkeley: University of California Press, 2013.

Pearse, C. Kenneth. "Grazing in the Middle East: Past, Present, and Future." *Journal of Range Management* 24, no. 1 (1971): 13–16.

Pecher, Yosefa. Interview by Tamar Novick. Kibbutz Mizra', February 9, 2012.

Penslar, Derek. *Zionism and Technocracy: The Engineering of Jewish Settlement in Palestine, 1870–1918*. Bloomington: Indiana University Press, 1991.

Perevolotsky, Avi, and No'am G. Seligman. "Role of Grazing in Mediterranean Rangeland Ecosystems." *Bioscience* 48 (1998): 1007–1017.

Pfeffer, Naomi. "Pioneers of Infertility Treatment." In *Women and Modern Medicine*, edited by in Lawrence Conrad and Ann Hardy, 245–261. Amsterdam: Editions Rodolpi B. V., 2001.

Pfeffer, Naomi. *The Stork and the Syringe: A Political History of Reproductive Medicine*. Cambridge, UK: Polity Press, 1993.

"Ph. J. Baldensperger." *Bee World* 29 (1948): 73.

Phillips, E. F. *Beekeeping*. Edited by Liberty Hyde Baily. New York: Macmillan Company, 1914.

Phillips, E. F. "Mr. P. J. Baldensperger." *Bee World* 8 (1927): 97–99.

Pickthall, Marmaduke W. *Oriental Encounters: Palestine and Syria, 1894-5-6*. London: W. Colloms Sons and Co., 1918.

Pinder, David. "Errant Paths: The Poetics and Politics of Walking." *Environment and Planning D: Society and Space* 29, no. 4 (2011): 672–692.

Pinhas, Shira. "Road, Map: Partition in Palestine from the Local to the Transnational." *Journal of Levantine Studies* 10, no. 1 (2020): 111–121.

Pinthus, Moshe Yoel. *History of Agricultural Research in Israel in the Pre-Statehood Era (1920–1949)*. Haifa: Itay Bahur, 2011.

Pitt, George. "Palestine." *British Friend* 40, no. 33 (1882): 257–258.

Plezental, Ayala. "'Milky Way': The Dairy Industry in Eretz Israel in the 1930s as a Mirror for German-Jewish Relations." In *Germany and Eretz Israel: Cultural Meeting Point*, edited by Moshe Zimmerman, 133–142. Jerusalem: Hebrew University Magnes Press, 2003.

Polanyi, Michael. *Personal Knowledge: Towards a Post-Critical Philosophy*. Chicago: University of Chicago Press, 1958.

Porcher, Jocelyne, and Jean Estebanez, eds. *Animal Labor: A New Perspective on Human-Animal Relations*. New York: Columbia University Press, 2020.

Porter, Theodore M. *Trust in Numbers: The Pursuit of Objectivity in Science and Public Life*. Princeton, NJ: Princeton University Press, 1995.

Potter, Captain Egon. "Ph. J. Baldensperger's Career." *Bee World* 8 (1927): 156–157.

Pratt, Geraldine, and Victoria Rosner, eds. *The Global and the Intimate: Feminism in Our Time*. New York: Columbia University Press, 2012.

Presner, Todd Samuel. *Muscular Judaism: The Jewish Body and the Politics of Regeneration*. London: Routledge, 2007.

Pri'el, Aharon. "'A Green Patrol' Will Operate against Bedouin Herds." *Ma'ariv*, August 1976, 6.

Pritchard, Sara B. *Confluence: The Nature of Technology and the Remaking of the Rhône*. Cambridge, MA: Harvard University Press, 2011.

Pritchard, Sara B. "Towards an Environmental History of Technology." In *The Oxford Handbook of Environmental History*, edited by Andrew C. Isenberg, 227–258. Oxford: Oxford University Press, 2014.

"Prof. Bernhard Zondek Died Two Days Ago in New-York." *Ma'ariv*, November 10, 1966, 2.

"The Project of Pinching Holes in Goat Ears and the Sale of Lands." *Filastin*, January 31, 1947, 4.

Rabinowitz, Dan. "Themes in the Economy of the Bedouin of South Sinai in the Nineteenth and Twentieth Centuries." *International Journal of Middle East Studies* 17, no. 2 (1985): 211–228.

Rader, Karen. *Making Mice: Standardizing Animals for American Biomedical Research.* Princeton, NJ: Princeton University Press, 2004.

Raffles, Hugh. "Intimate Knowledge." *International Social Science Journal* 173 (2000): 325–334.

Raviv, Yael. "Falafel: A National Icon." *Gastronomica: The Journal of Critical Food Studies* 3, no. 3 (2003): 20–25.

Reichert, Israel. "Yizthak Wilkansky—for His 50th Birthday." *Davar*, February 2, 1931, 2.

Reichert, Israel. "To Yzhak Elazari-Volcani on His 70th Birthday: The Architect of the Settlement and Labor Zionism." *Davar*, February 3, 1950.

Ritvo, Hariet. *The Animal Estate: The English and Other Creatures in the Victorian Age.* Cambridge, MA: Harvard University Press, 1987.

Ritvo, Hariet. *Noble Cows and Hybrid Zebras: Essays on Animals and History.* Charlottesville: University of Virginia Press, 2010.

Robinson, Shira. *Citizen Strangers: Palestinians and the Birth of Israel's Liberal Settler State.* Stanford, CA: Stanford University Press, 2013.

Rosenberg, Gabriel N. "No Scrubs: Livestock Breeding, Eugenics, and the State in Early Twentieth-Century United States." *Journal of American History* 107, no. 2 (2020): 362–387.

Rosenberg, Miriam, Yeshayahu Folman, Zeev Herz, Israel Flamenbaum, Amiel Berman, and Moshe Kaim. "Effect of Climatic Conditions on Peripheral Concentrations of LH, Progesterone and Oestradiol-17b in High Milk-Yielding Cows." *Journal of Reproduction and Fertility* 66, no. 1 (1982): 139–146.

Rubin, Reuven. *Self-Portrait with a Goat.* 1924. http://www.bridgemanimages.com /en-GB/asset/825853/rubin-reuven-1893-1974/self-portrait-1924-oil-on-canvas-on -masonite.

Rueber, Micah Aaron. "Making Milking Modern: Agriculture Science and the American Dairy, 1890–1940." PhD diss., Mississippi State University, 2010.

Russell, Nicholas. *Like Engend'ring Like: Heredity and Animal Breeding in Early Modern England.* Cambridge: Cambridge University Press, 1986.

Sabbagh-Khouri, Areej. "Tracing Settler Colonialism: A Genealogy of a Paradigm in the Sociology of Knowledge Production in Israel." *Politics & Society* 50, no. 1 (2022): 44–83.

Sadan, Michal. *The Hebrew Shepherd: Transformation of Image and Symbol from the Hebrew Enlightenment Literature to the New Hebrew Culture in Israel.* Jerusalem: Yad Ben-Zvi, 2011.

Safran, Yair, and Tamir Goren. "Ideas and Plans to Construct a Railroad in Northern Palestine in the Later Ottoman Period." *Middle Eastern Studies* 46, no. 5 (2010): 753–770.

Saha, Jonathan. "Milk to Mandalay: Dairy Consumption, Animal History, and the Political Geography of Colonial Burma." *Journal of Historical Geography* 54 (2016): 1–12.

Said, Edward. *Orientalism*. New York: Pantheon Books, 1978.

Sallah Shabati. Directed by Ephraim Kishon. Produced by Menahem Golan. 1964.

Sallon, Sarah, Emira Cherif, Nathalie Chabrillange, Elaine Solowey, Muriel Gros-Balthazard, Sarah Ivor-Ra, Jean-Frédéric Terral, et al. "Origins and Insights into the Historic Judean Date Palm Based on Genetic Analysis of Germinated Ancient Seeds and Morphometric Studies." *Science Advances* 6, no. 6 (2020): 1–10.

Sallon, Sarah, and Elaine Solowey. "Germination, Genetics, and Growth of an Ancient Date Seed." *Science* 320 (June 2008): 1464.

Salzmann, Ariel. "An Ancien Régime Revisited: 'Privatization' and Political Economy in the Eighteenth-Century Ottoman Empire." *Politics and Society* 21, no. 4 (1993): 393–4323.

Saraiva, Tiago. *Fascist Pigs: Technoscientific Organisms and the History of Fascism*. Cambridge, MA: MIT Press, 2018.

Satia, Priya. "'A Rebellion of Technology': Development, Policing, and the British Arabian Imaginary." In *Environmental Imaginaries of the Middle East and North Africa*, edited by Diana K. Davis and Edmund Burke III, 23–59. Athens: Ohio University Press, 2011.

Satia, Priya. *Spies in Arabia: The Great War and the Cultural Foundations of Britain's Covert Empire in the Middle East*. Oxford: Oxford University Press, 2008.

Sat-Ky, Moshe. "A Short Guide for Raising Dairy Cattle." *Hasade* 17 (1937): 441–447.

Sauer, Paul. *The Holy Land Called: The Story of the the Temple Society*. Stuttgart: Konrad Theiss Verlag GmbH, 1985.

Schaff, Philip. *Through the Land of the Bible: Egypt, the Desert, and Palestine*. New York: American Tract Society, 1878.

Schiff, Benjamin N. *Refugees unto the Third Generation: UN Aid to Palestinians*. Syracuse, NY: Syracuse University Press, 1995.

Schölch, Alexander. "The Economic Development of Palestine, 1856–1882." *Journal of Palestine Studies* 10, no. 3 (1981): 35–58.

Schorr, Ag. Moshe. *The House Goat*. Tel Aviv: Hakarmel [published by the chief supervisor of agricultural education, Ministry of Education and Culture], 1949.

Scott, James C. *The Art of Not Being Governed: An Anarchist History of Upland Southeast Asia*. New Haven, CT: Yale University Press, 2009.

Scott, James C. *The Moral Economy of the Peasant: Rebellion and Subsistence in Southeast Asia*. New Haven, CT: Yale University Press, 1976.

Scott, James C. *Seeing Like a State: How Certain Schemes to Improve the Human Condition Have Failed*. New Haven, CT: Yale University Press, 1998.

Scott, James C. *Weapons of the Weak: Everyday Forms of Peasant Resistance*. New Haven, CT: Yale University Press, 1985.

Şen, Gül. "The Landscape of Southern Bilād al-Shām through the Eyes of the Sixteenth-Century Ottoman Cosmographer Āşik Mehmed." In *Living with Nature and Things: Contribution to a New Social History of the Middle Islamic Periods*, edited by Bethany J. Walker and Abdelkader Al Ghouz, 49–78. Bonn: Bonn University Press, 2020.

"'Senior Home' for Stavit." *Davar*, July 2, 1951, 4.

"A Severe Dispute between the Druze and the Ministry of Agriculture regarding the Grazing of Goats." *Ha'aretz*, January 24, 1954.

Shafir, Gershon. *Land, Labor and the Origins of the Israeli-Palestinian Conflict, 1882–1914*. Cambridge: Cambridge University Press, 1989.

Shalev, Meir. *A Pigeon and a Boy: A Novel*. Tel Aviv: Am Oved, 2006.

Shapin, Steven, and Simon Schaffer. *Leviathan and the Air-Pump: Hobbes, Boyle, and the Experimental Life*. Princeton, NJ: Princeton University Press, 1985.

Sharkey, Heather J. *American Evangelicals in Egypt: Missionary Encounters in an Age of Empire*. Princeton, NJ: Princeton University Press, 2008.

Sharoni, G. "The Goats of Israel Assembled in the College." *Ma'ariv*, August 30, 1953, 2.

Shavit, Yaacov. *From Hebrew to Canaanite: Aspects in the History, Ideology and Utopia of the "Hebrew Renaissance"—From Radical Zionism to Anti-Zionism*. Jerusalem: Domino Press, 1984.

Shavit, Yaacov, and Mordechai Eran. *The Hebrew Bible Reborn: From Holy Scripture to the Book of Books, a History of Biblical Culture and the Battles over the Bible in Modern Judaism*. Berlin: Walter de Gruyter, 2007.

Shavit, Yaacov, and Dan Gil'adi. "The Cowshed and the Agricultural Economy in Eretz Israel: The Place and Role of the Dairy Farming in the Jewish Settlement Program in Eretz Israel during the Mandate Period." *Catedra* 18 (1981): 178–193.

Shehadeh, Raja. *Palestinian Walks: Notes on a Vanishing Landscape*. London: Profile, 2007.

Shelem, Matityahu. "Know Dear Shepherd." *Zemereshet*, May 9, 2012. http://www.zemereshet.co.il/song.asp?id=654.

Shelem, Matityahu. "The Sheep Have Spread." *Zemereshet*, May 9, 2012. http://www .zemereshet.co.il/song.asp?id=1688.

Shohet, Dror, and Dan Shohet. "Our Camels Are Part of the Desert." 2012. https:// www.facebook.com/notes/dror-shohet/%D7%94%D7%92%D7%9E%D7%9C% D7%99%D7%9D-%D7%A9%D7%9C%D7%A0%D7%95-%D7%94%D7%9D-% D7%97%D7%9C%D7%A7-%D7%9E%D7%94%D7%9E%D7%93%D7%91%D7 %A8/10151891597981866.

Siegel-Itzkovitch, Judy. "Medicinal Date Palm from Oldest Known Seed Planted." *Jerusalem Post*, November 25, 2011. http://www.jpost.com/Health/Article.aspx?id =246956.

Simpson, Sir John Hope. *Palestine: Report on Immigration, Land Settlement and Development*. London: Secretary of State for the Colonies, 1930.

Sinai, Smadar. *Miriam Baratz—Portrait of a Pioneer*. Ramat-Ef'al: Yad Tebenkin, 2002.

Singer, Amy. *Palestinian Peasants and Ottoman Officials: Rural Administration around Sixteenth-Century Jerusalem*. Cambridge: Cambridge University Press, 1994.

Singer, Irma. "First Dairywoman of the First 'Kvutza.'" *Palestine Post*, February 15, 1938.

Singer, Isaac Bashevis. *Zlateh the Goat and Other Stories*. New York: HarperCollins, 1966.

Sivaramakrishnan, Kalyanakrishnan. "Science, Environment and Empire History: Comparative Perspectives from Forests in Colonial India." *Environment and History* 14 (2008): 41–65.

Skoss, Solomon Leon. *Portrait of a Jewish Scholar: Essays and Addresses*. New York: Block Publishing Company, 1957.

Smith, J. M., and S. J. Gilbert. "Cattle Disease Occurring in Palestine and Segregation and Prevention Control of Bovine Contagious Abortion." *Journal of Comparative Pathology and Therapeutics* 47 (1934): 94–106.

Smith, Jenny Leigh. *Works in Progress: Plans and Realities on Soviet Farms, 1930–1963*. New Haven, CT: Yale University Press, 2014.

Smith-Howard, Kendra. *Pure and Modern Milk: An Environmental History since 1900*. Oxford: Oxford University Press, 2014.

Solowey, Elaine. Interview by Tamar Novick. Kibbutz Ketura, February 16, 2012.

Soto Laveaga, Gabriela. *Jungle Laboratories: Mexican Peasants, National Projects, and the Making of the Pill*. Durham, NC: Duke University Press, 2009.

"Sperm of Bulls Arrived from the United-States by Plane." *Al-Hamishmar*, February 10, 1947, 4.

Spoer, A. M. "Palestine Folktales." *Folklore* 42, no. 2 (1931): 150–156.

Stead, K. W. *Report on the Economic and Financial Situation of Palestine.* London: H. M. Stationery Office, 1927.

Stephan, Stephan Hanna. "Palestinian Animal Stories and Fables." *Journal of the Palestine Oriental Society* 3 (1923): 167–190.

"Sterility in Women Becoming a Serious Problem: Professor Zondek." *Winnipeg Tribune,* September 3, 1952.

Stern. "Finding Out a Method." *Hasade* 5 (1925).

Stolberg, Michael. "The Decline of Uroscopy in Early Modern Learned Medicine, 1500–1650." *Early Science and Medicine* 12 (2007): 313–336.

Stoler, Ann Laura. *Along the Archival Grain: Epistemic Anxieties and Colonial Common Sense.* Princeton, NJ: Princeton University Press, 2010.

Stoler, Ann Laura. *Carnal Knowledge and Imperial Power: Race and the Intimate in Colonial Rule.* Berkeley: University of California Press, 2002.

Stoler, Ann Laura. "Colonial Aphasia: Race and Disabled Histories in France." *Public Culture* 23, no. 1 (2011): 121–156.

Stoller, Shmuel. *The Return of the Date Palm to the Land of Israel: Ben-Zion Israeli's Journeys in Iraq, Iran, and Egypt.* Tel Aviv: Hakibbutz Hameuchad Publishing House, 1977.

Stone, Gillian. "Gilbert Noel Sale, 1897–1991." *Commonwealth Forestry Review* 70, no. 3 (1991): 88–90.

Strange, James P. "A Severe Stinging and Much Fatigue—Frank Benton and His 1881 Search for Apis Dorsata." *American Entomologist* 47, no. 2 (2001): 112–116.

Sufian, Sandra M. *Healing the Land and the Nation: Malaria and the Zionist Project in Palestine, 1920–1947.* Chicago: University of Chicago Press, 2007.

Sulman, Felix Gad. *Health, Weather and Climate.* Basel: S. Krager, 1976.

Sulman, Felix Gad. *Hormones: Textbook for Hormone Theory and Use.* Tel Aviv: Hasade Library, 1962.

Sulman, Felix Gad. *Hypothalamic Control of Lactation: Monographs on Endocrinology.* London: William Heinemann Medical Books, 1970.

Sulman, Felix Gad, and Edith Sulman. "Pregnancy Test with the Male Toad." *An International Journal of Obstetrics and Gynecology* 56, no. 6 (1949): 1014–1017.

Sulman, Felix Gad, and Gavriella Tietz. "The Sex-Hormones of the Mandrake." *Harefua—Journal of the Palestine Jewish Medical Association* 23, no. 7 (1942): 1–4.

Swedenburg, Ted. *Memories of Revolt: The 1936–1939 Rebellion and the Palestinian National Past.* Fayetteville: University of Arkansas Press, 2003.

Swyngedouw, Erik. *Liquid Power: Contested Hydro-Modernities in Twentieth-Century Spain.* Cambridge, MA: MIT Press, 2015.

Tabor, Vivien. "He Maketh the Barren Woman . . . A Joyful Mother of Children." *Hadassah Newsletter*, April 1960.

Tal, Alon. *Pollution in a Promised Land: An Environmental History of Israel.* Berkeley: University of California Press, 2002.

"Talk of the Fellah: There Is No Milk Left in This Cow." *Mir'at Al-Sharq*, November 22, 1930, 1.

Talmon, Naftali. "The Agricultural Farm in the Templar Settlements and Its Contribution for the Development of Agriculture in Eretz Israel." *Catedra* 78 (1995): 65–81.

Talmon, Naftali. "Fritz Keller—Of the Pioneers of Modern Agriculture in Eretz Israel." In *The Landscape of His Homeland: Studies in the Geography and History of Eretz Israel (Dedicated to Yehosuah Ben-Arieh)*, edited by Yossi Ben-Artzi, Israel Bartal, and Elhana Riner, 333–351. Jerusalem: Hebrew University Magnes Press, 1999.

Tamari, Salim. *Mountain against the Sea: Essays on Palestinian Society and Culture.* Berkeley: University of California Press, 2009.

Tausk, Marius. *Organon: The Story of an Unusual Pharmaceutical Enterprise.* Oss: Akzo Pharma bv, 1984.

Taylor, Charles. *A Secular Age.* Cambridge, MA: Harvard University Press, 2007.

Teman, Elly. *Birthing the Mother: The Surrogate Body and the Pregnant Self.* Berkeley: University of California Press, 2010.

Tesdell, Omar Loren. "Shadow Spaces: Territory, Sovereignty, and the Question of Palestinian Cultivation." PhD diss., University of Minnesota, 2013.

Tezcan, Baki. *The Second Ottoman Empire: Political and Social Transformation in the Early Modern World.* Cambridge: Cambridge University Press, 2010.

Theunissen, Bert. "Breeding for Nobility or for Production? Cultures of Dairy Cattle Breeding in the Netherlands, 1945–1995." *Isis* 103, no. 2 (2012): 278–309.

Thompson, E. P. "Time, Work-Discipline, and Industrial Capitalism." *Past and Present* 38 (1967): 56–97.

Thomson, William McClure. *The Land and the Book; or, Biblical Illustrations Drawn from the Manners and Customs, the Scenes and Scenery, of the Holy Land.* London: T. Nelson and Sons, 1859.

Tilley, Helen. *Africa as a Living Laboratory: Empire, Development, and the Problem of Scientific Knowledge, 1870–1950.* Chicago: University of Chicago Press, 2011.

Tirosh, David. *The Emek Train.* Tel Aviv: Society for the Protection of Nature, 1988.

Toshek: Avraham Amarent. Tel Aviv: Sifriyat Poalim, 1993.

Tresch, John. *The Romantic Machine: Utopian Science and Technology after Napoleon.* Chicago: University of Chicago Press, 2012.

Troen, Ilan S. *Imagining Zion: Dreams, Designs, and Realities in a Century of Jewish Settlement.* New Haven, CT: Yale University Press, 2003.

University reporter. "An Award for a Praised Researcher for His Scientific Achievements." *Davar*, April 24, 1957.

Uri, Y. "For the Memory of Yitzhak Volcani (a Year after His Death)." *Davar*, May 25, 1956, 5.

Valencius, Conevery Bolton. *The Health of the Country: How American Settlers Understood Themselves and Their Land.* New York: Basic Books, 2002.

Valeze, Deaborah. *Milk: A Local and Global History.* New Haven, CT: Yale University Press, 2011.

Verlinsky, Nahum. *Debating Production.* Tel Aviv: Tnuva Center, 1973.

Vogel, Lester I. *To See Promised Land: Americans and the Holy Land in the Nineteenth Century.* University Park: Penn State University Press, 1993.

Volcani, Professor Ra'anan. Interview by Nir Mann. Rehovot, 1996.

Wachs, Elizabeth, and Alon Tal. "Herd No More: Livestock Husbandry Policies and the Environment in Israel." *Journal of Environmental Ethics* 22 (2009): 401–422.

Wacquant, Loïc J. D. "Pugs at Work: Bodily Capital and Bodily Labor among Professional Boxers." *Body and Society* 1 (1995): 65–93.

Wade, Nicholas. "Dead for 32,000 Years, an Arctic Plant Is Revived." *New York Times*, February 12, 2012.

The Wanted 18. Directed by Amer Shomali and Paul Cowan. 2014.

Warren, Louis S. *The Hunter's Game: Poachers and Conservationists in Twentieth-Century America.* New Haven, CT: Yale University Press, 1997.

Weber, Max. *The Protestant Ethic and the Spirit of Capitalism.* London: Routledge, 1992. First published 1905.

Webster, Charles. *The Great Instauration: Science, Medicine, and Reform, 1626–1660.* London: Duckworth, 1975.

Weinstock, Tikva. "Miracle Drug for Sterile Women—In Elderly Homes." *Ma'ariv*, February 1, 1965, 9.

Weiss, Meira. *The Chosen Body: The Politics of the Body in Israeli Society.* Stanford, CA: Stanford University Press, 2002.

White, Richard. *The Organic Machine: The Remaking of the Columbia River.* New York: Hill and Wang, 1995.

White, Sam. *The Climate of Rebellion in the Early Modern Ottoman Empire.* New York: Cambridge University Press, 2011.

Wick, Livia. "Making Lives under Closure: Birth and Medicine in Palestine's Waiting Zones." PhD diss., Massachusetts Institute of Technology, 2006.

"Will Palestine Become a Center for Cow Breeding?" *Haqiqat al-Amr*, July 1939, 2.

Williams, Elizabeth R. "Cultivating Empires: Environment, Expertise, and Scientific Agriculture in Late Ottoman and French Mandate Syria." PhD diss., Georgetown University, 2015.

Wilmot, Sara. "From 'Public Service' to Artificial Insemination: Animal Breeding Science and Reproductive Research in Early Twentieth-Century Britain." *Studies in History and Philosophy of Biological and Biomedical Sciences* 38 (2007): 411–441.

Wishnitzer, Avner. *Reading Clocks, Alla Turca: Time and Society in the Late Ottoman Empire*. Chicago: Chicago University Press, 2015.

Wohl, Anthony S. "'Ben JuJu': Representations of Disraeli's Jewishness in the Victorian Political Cartoon." *Jewish History* 10, no. 2 (1996): 89–134.

Wolfenson, D., William W. Thatcher, Lokenga Badinga, J. D. Savio, Rina Meidan, B. J. Lew, Ruth Braw-Tal, and Amiel Berman. "Effect of Heat Stress on Follicular Development during the Estrous Cycle in Lactating Dairy Cattle." *Biology of Reproduction* 52, no. 5 (1995): 1106–1113.

Woods, Abigail. "The Farm as a Clinic: Veterinary Expertise and the Transformation of Dairy Farming, 1930–1950." *Studies in History and Philosophy of Biological and Biomedical Sciences* 38 (2007): 462–487.

Woods, Rebecca J. H. *The Herds Shot Round the World: Native Breeds and the British Empire, 1800–1900*. Chapel Hill: University of North Carolina Press, 2017.

Worster, Donald. *Rivers of Empire: Water, Aridity, and the Growth of the American West*. New York: Oxford University Press, 1985.

Yaron, Hadas. *Zionist Arabesques: Modern Landscapes, Non-Modern Texts*. Boston: Academic Studies Press, 2010.

Yazbak, Mahmoud. "From Poverty to Revolt: Economic Factors in the Outbreak of the 1936 Rebellion in Palestine." *Middle Eastern Studies* 36, no. 3 (2000): 93–113.

Zagorodsky, Elimelech. *The Book of Milk, Part I: Milk and Its* Outcomes. Jaffa: Hachak'lai Publishers, 1914.

Zagorodsky, Elimelech. "Milk and Its Outcomes," *Hameshek Hahaklai* 3, no. 3 (1914): 65-67.

Zakkariyah, Wasfi Beck. "Cattle Raising." *Al-Iqtisadat al-'Arabiyya*, December 2, 1936, 8–9.

Zamir, David, and Matityahu Shelem, eds. *The Hebrew Shepherd*. Merhavia: Hebrew Shepherds Association, 1957.

Zeira, Motti. *Man of Loves: The Story of Yehoshua Brandstetter.* Jerusalem: Yad Ben-Zvi, 2006.

Zerubavel, Yael. "Memory, the Rebirth of the Native, and the 'Hebrew Bedouin' Identity." *Social Research* 75, no. 1 (2008): 315–352.

Zerubavel, Yael. *Recovered Roots: Collective Memory and the Making of Israeli National Tradition.* Chicago: University of Chicago Press, 1995.

Zierer, Clifford M. "Migratory Beekeepers of Southern California." *Geographical Review* 22, no. 2 (1932): 260–269.

Zikri, Ben. "Only a Third of Israeli Camels Have ID Chips—and It Could Be Deadly." *Ha'aretz,* October 7, 2019. https://www.haaretz.com/israel-news/.premium-only-a -third-of-bedouin-camels-in-israel-s-south-have-id-chips-1.7949607.

Zondek, Bernhard. *Estrous Hormone and a Colored Matter Resembling Vitamin B2 in the Dead Sea.* Jerusalem: Magnes Books, 1938.

Zondek, Bernhard. "Impairment of Anterior Pituitary Functions by Follicular Hormone." *Lancet* (October 10, 1936): 842–847.

Zondek, Bernhard. "Mass Excretion of Oestrogenic Hormone in the Urine of Stallions." *Nature* (February 10, 1934): 209–210.

Zondek, Bernhard. "On the Mechanism of the Action of Gonadotropin from Pregnancy Urine." *Journal of Endocrinology* 2 (1940): 12–20.

Zondek, Bernhard. "Oestrogenic Hormone in the Urine of the Stallion." *Nature* (March 31, 1934): 494.

Zondek, Bernhard. "Oestrogenic Substances in the Dead Sea." *Nature* 140, no. 3536 (1937): 240.

Zondek, Bernhard. "Studies of the Mechanism of the Female Genital System." *Fertility and Sterility* 10, no. 1 (1959): 12.

Zondek, Bernhard, and Rivka Black. "Is Normal Human Urine Toxic?" *Proceedings of the Society for Experimental Biology and Medicine* 61 (1946): 140–142.

Zondek, Bernhard, and Y. M. Bromberg. "Endocrine Allergy: Clinical Reaction of Allergy to Endogenous Hormones and Their Treatment." *Journal of Obstetrics and Gynecology of the British Empire* 54, no. 1 (1947): 1–19.

Zondek, Bernhard, and Felix Sulman. "The Antigonadotropic Factor: Species Specificity and Organ Specificity." *Proceedings of the Society for Experimental Biology and Medicine* 36 (1937): 712–717.

Zondek, Bernhard, and Felix Sulman. *The Antigonadotropic Factor, with Consideration of the Antihormone Problem.* Baltimore: Williams and Wilkins Company, 1942.

Zondek, Bernhard, and Felix Sulman. "The Role of Sex Hormones in the Sex System of the Animals (Lecture Given at the Jewish Organization of Veterinary

Services Annual Conference)." *Alon Histadrut Harofim Haveterinarim Beeretz Israel* 5 (1940): 8.

Zondek, Bernhard, and Felix Sulman. "A Twenty-Four Hour Pregnancy Test for Equines." *Nature* 455 (March 10, 1945): 320.

Zu'bi, Himmat. "Palestinian Fertility in the Israeli Sphere: Palestinian Women in Israel Undergoing IVF Treatments." In *Bioethics and Biopolitics in Israel: Socio-Legal, Political, and Empirical Analysis*, edited by Hagai Boas, Yael Hashiloni-Dolev, Nadav Davidovitch, Dani Filk, and Shai J. Lavi, 160–180. Cambridge: Cambridge University Press, 2018.

INDEX

Mikael Hård and Thomas J. Misa, editors, *Urban Machinery: Inside Modern European Cities*

Christine Hine, *Systematics as Cyberscience: Computers, Change, and Continuity in Science*

Wesley Shrum, Joel Genuth, and Ivan Chompalov, *Structures of Scientific Collaboration*

Shobita Parthasarathy, *Building Genetic Medicine: Breast Cancer, Technology, and the Comparative Politics of Health Care*

Kristen Haring, *Ham Radio's Technical Culture*

Atsushi Akera, *Calculating a Natural World: Scientists, Engineers, and Computers during the Rise of U.S. Cold War Research*

Donald MacKenzie, *An Engine, Not a Camera: How Financial Models Shape Markets*

Geoffrey C. Bowker, *Memory Practices in the Sciences*

Christophe Lécuyer, *Making Silicon Valley: Innovation and the Growth of High Tech, 1930–1970*

Anique Hommels, *Unbuilding Cities: Obduracy in Urban Sociotechnical Change*

David Kaiser, editor, *Pedagogy and the Practice of Science: Historical and Contemporary Perspectives*

Charis Thompson, *Making Parents: The Ontological Choreography of Reproductive Technology*

Pablo J. Boczkowski, *Digitizing the News: Innovation in Online Newspapers*

Dominique Vinck, editor, *Everyday Engineering: An Ethnography of Design and Innovation*

Nelly Oudshoorn and Trevor Pinch, editors, *How Users Matter: The Co-Construction of Users and Technology*

Peter Keating and Alberto Cambrosio, *Biomedical Platforms: Realigning the Normal and the Pathological in Late-Twentieth-Century Medicine*

Paul Rosen, *Framing Production: Technology, Culture, and Change in the British Bicycle Industry*

Maggie Mort, *Building the Trident Network: A Study of the Enrollment of People, Knowledge, and Machines*

Donald MacKenzie, *Mechanizing Proof: Computing, Risk, and Trust*

Geoffrey C. Bowker and Susan Leigh Star, *Sorting Things Out: Classification and Its Consequences*

Charles Bazerman, *The Languages of Edison's Light*

Janet Abbate, *Inventing the Internet*

Herbert Gottweis, *Governing Molecules: The Discursive Politics of Genetic Engineering in Europe and the United States*

Kathryn Henderson, *On Line and on Paper: Visual Representation, Visual Culture, and Computer Graphics in Design Engineering*

Susanne K. Schmidt and Raymund Werle, *Coordinating Technology: Studies in the International Standardization of Telecommunications*

Marc Berg, *Rationalizing Medical Work: Decision Support Techniques and Medical Practices*

Eda Kranakis, *Constructing a Bridge: An Exploration of Engineering Culture, Design, and Research in Nineteenth-Century France and America*

Paul N. Edwards, *The Closed World: Computers and the Politics of Discourse in Cold War America*

Donald MacKenzie, *Knowing Machines: Essays on Technical Change*

Wiebe E. Bijker, *Of Bicycles, Bakelites, and Bulbs: Toward a Theory of Sociotechnical Change*

Louis L. Bucciarelli, *Designing Engineers*

Geoffrey C. Bowker, *Science on the Run: Information Management and Industrial Geophysics at Schlumberger, 1920–1940*

Wiebe E. Bijker and John Law, editors, *Shaping Technology / Building Society: Studies in Sociotechnical Change*

Stuart Blume, *Insight and Industry: On the Dynamics of Technological Change in Medicine*

Donald MacKenzie, *Inventing Accuracy: A Historical Sociology of Nuclear Missile Guidance*

Pamela E. Mack, *Viewing the Earth: The Social Construction of the Landsat Satellite System*

H. M. Collins, *Artificial Experts: Social Knowledge and Intelligent Machines*

http://mitpress.mit.edu/books/series/inside-technology